# Robots,
# Androids, and
# Animatrons

# Robots, Androids, and Animatrons

## 12 Incredible Projects You Can Build

*John Iovine*

Second Edition

**McGraw-Hill**

New York   Chicago   San Francisco   Lisbon   London   Madrid
Mexico City   Milan   New Delhi   San Juan   Seoul
Singapore   Sydney   Toronto

**Library of Congress Cataloging-in-Publication Data**

Iovine, John.
    Robots, androids, and animatrons : 12 incredible projects you can
build / John Iovine.—2nd ed.
       p.    cm.
    Includes index.
    ISBN 0-07-137683-6
    1. Robotics—Amateurs' manuals.  I. Title.
TJ211.5.I58    2001
629.8'92—dc21
                                                2001044821

# *McGraw-Hill*

*A Division of The **McGraw·Hill** Companies*

2 3 4 5 6 7 8 9 0  DOC/DOC  0 7 6 5 4 3 2

ISBN 0-07-137683-6

*The sponsoring editor for this book was Scott L. Grillo, the editing
supervisor was Stephen M. Smith, and the production supervisor was
Pamela A. Pelton. It was set in ITC Century Light following the EL3
design by Wayne A. Palmer of McGraw-Hill Professional's Hightstown,
N.J., composition unit.*

*Printed and bound by R. R. Donnelley & Sons Company.*

McGraw-Hill books are available at special quantity discounts to use as
premiums and sales promotions, or for use in corporate training programs. For
more information, please write to the Director of Special Sales, McGraw-Hill
Professional, Two Penn Plaza, New York, NY 10121-2298. Or contact your local
bookstore.

 This book is printed on recycled, acid-free paper containing a
minimum of 50 percent recycled, de-inked fiber.

Dedication:
To Ellen, my wife;
James, my son; and
AnnaRose, my daughter—
with love

# Contents

## 15 Robotic arm and IBM PC interface and speech control   281

## 16 Android hand   299

# Introduction

There are many interesting and fun things to do in electronics, and one of the most enjoyable is building robots. Not only do you employ electronic circuits and systems, but they must be merged with other technologies. Building a robot from scratch involves the following:

☐ Power supply systems

☐ Motors and gears for drive and motion control

☐ Sensors

☐ Artificial intelligence

Each one of these technologies has numerous books dedicated to its study. Naturally, a comprehensive look at each technology isn't possible in one book, but we will touch upon these areas, and you will gain hands-on knowledge and a springboard for future experimentation.

Robotics is an evolving technology. There are many approaches to building robots, and no one can be sure which method or technology will be used 100 years from now. Like biological systems, robotics is evolving following the Darwinian model of survival of the fittest.

You're not alone when you become a robotist. I was surprised to learn that there are many people, government organizations, private organizations, competitions, and clubs devoted to the subject of amateur robotics. NASA has the most advanced robotics systems program I ever saw. Much of the information is free for the asking. If you have Internet access, jump to one of the search engines (Yahoo, Excite, etc.) and search under robotics. You will find the websites of many companies, individuals, universities, clubs, and newsgroups dedicated to robotics.

# Acknowledgments

I would like to thank some of the people who helped make this book possible: Matt Wagner, my agent at Waterside Productions; Scott Grillo, who tried to keep me on schedule; and Stephen Smith for a great job of editing.

# In the beginning

SOME HISTORIANS BELIEVE THE ORIGIN OF ROBOTICS CAN be traced back to the ancient Greeks. It was around 270 BC when Ctesibus (a Greek engineer) made organs and water clocks with movable figures.

Other historians believe robotics began with mechanical dolls. In the 1770s, Pierre Jacquet-Droz, a Swiss clock maker and inventor of the wristwatch, created three ingenious mechanical dolls. He made the dolls so that each one could perform a specific function: one would write, another would play music on an organ, and the third could draw a picture. As sophisticated as they were, the dolls, whose purpose was to amuse royalty, performed all their respective feats using gears, cogs, pegs, and springs.

More recently, in 1898, Nikola Tesla built a radio-controlled submersible boat. This was no small feat in 1898. The submersible was demonstrated in Madison Square Garden. Although Nikola Tesla had plans to make the boat autonomous, lack of funding prevented further research.

The word "robot" was first used in a 1921 play titled *R.U.R.: Rossum's Universal Robots*, by Czechoslovakian writer Karel Capek. *Robot* is a Czech word meaning "worker." The play described mechanical servants, the "robots." When the robots were endowed with emotion, they turned on their masters and destroyed them.

Historically, we have sought to endow inanimate objects that resemble the human form with human abilities and attributes. From this is derived the word *anthrobots,* robots in human form.

Since Karel Capek's play, robots have become a staple in many science fiction stories and movies. As robots evolved, so did the terminology needed to describe the different robotic forms. So, in addition to the old "tin-man" robot, we also have *cyborgs,* which are part human and part machine, and *androids,* which are specially built robots designed to be humanlike.

Many people had their first look at a real robot during the 1939 World's Fair. Westinghouse Electric built a robot they called Elektro the Moto Man. Although Elektro had motors and gears to move its mouth, arms, and hands, it could not perform any useful work. It was joined on stage by a mechanical dog named Sparko.

## Why build robots?

Robots are indispensable in many manufacturing industries. The reason is that the cost per hour to operate a robot is a fraction of the cost of the human labor needed to perform the same function. More than this, once programmed, robots repeatedly perform functions with a high accuracy that surpasses that of the most experienced human operator. Human operators are, however, far more versatile. Humans can switch job tasks easily. Robots are built and programmed to be job specific. You wouldn't be able to program a welding robot to start counting parts in a bin.

Today's most advanced industrial robots will soon become "dinosaurs." Robots are in the infancy stage of their evolution. As robots evolve, they will become more versatile, emulating the human capacity and ability to switch job tasks easily.

While the personal computer has made an indelible mark on society, the personal robot hasn't made an appearance. Obviously there's more to a personal robot than a personal computer. Robots require a combination of elements to be effective: sophistication of intelligence, movement, mobility, navigation, and purpose.

## Purpose of robots

In the beginning, personal robots will focus on a singular function (job task) or purpose. For instance, today there are small mobile robots that can autonomously maintain a lawn by cutting the grass. These robots are solar powered and don't require any training. Underground wires are placed around the lawn perimeter. The robots sense the wires, remain within the defined perimeter, and don't wander off.

Building a useful personal robot is very difficult. In fact it's beyond the scope of this book, or for that matter, every other contemporary book on robotics. So you may reasonably ask, "What's the purpose of this book?" Well, in reading this book and building a few robots you gain entry into and become part of the ongoing robotic evolution.

Creativity and innovation do not belong to only those with college degrees. Robot building is not restricted to Ph.D.s, professors, universities, and industrial companies. By playing and experimenting with robots you can learn many aspects of robotics: artificial intelligence, neural networks, usefulness and purpose, sensors, navigation, articulated limbs, etc. The potential is to learn first hand about robotics and possibly make a contribution to the existing body of knowledge on robotics. And to this end amateur robotists do contribute, in some cases creating a clever design that surpasses mainstream robotic development.

As the saying goes, look before you leap. The first question to ask yourself when beginning a robot design is, "What is the purpose of this robot? What will it do and how will it accomplish its task?" My dream is to build a small robot that will change my cat's litter box.

This book provides the necessary information about circuits, sensors, drive systems, neural nets, and microcontrollers for you to build a robot. But before we begin, let's first look at a few current applications and how robots may be used in the future. The National Aeronautics and Space Administration (NASA) and the U.S. military build the most sophisticated robots. NASA's main interest in robotics involves (couldn't you guess) space exploration and telepresence. The military on the other hand utilizes the technology in warfare.

## Exploration

NASA routinely sends unmanned robotic explorers where it is impossible to send human explorers. Why send robots instead of humans? In a word, economics. It's much cheaper to send an expendable robot than a human. Humans require an enormous support system to travel into space: breathable atmosphere, food, heat, and living quarters. And, quite frankly, most humans would want to live through the experience and return to Earth in their lifetime.

Explorer spacecraft travel through the solar system where their electronic eyes transmit back to Earth fascinating pictures of the planets and their moons. The Viking probes sent to Mars looked for life and sent back pictures of the Martian landscape. NASA is developing planetary rovers, space probes, spider-legged walking explorers, and underwater rovers. NASA has the most advanced telerobotic program in the world, operating under the Office of Space Access and Technology (OSAT).

NASA estimates that by the year 2004, 50 percent of extra vehicle activity (EVA) will be conducted using telerobotics. For a complete explanation of telerobotics and telepresence, see Chap. 9.

Robotic space probes launched from Earth have provided spectacular views of our neighboring planets in the solar system. And in this era of tightening budgets, robotic explorers provide the best value for the taxpayer dollar. Robotic explorer systems can be built and implemented for a fraction of the cost of manned flights. Let's examine one case. The Mars Pathfinder represents a new generation of small, low-cost spacecraft and explorers.

### Mars Pathfinder (Sojourner)

The Mars Pathfinder consists of a lander and rover. It was launched from Earth in December of 1996 on board a McDonnell Douglas Delta II rocket and began its journey to Mars. It arrived on Mars on July 4, 1997.

The Pathfinder did not go into orbit around Mars; instead it flew directly into Mars's atmosphere at 17,000 miles per hour (mph) [27,000 kilometers per hour (km/h) or 7.6 kilometers per second (km/s)]. To prevent Pathfinder from burning up in the atmosphere, a combination of a heat shield, parachute, rockets, and airbags was used. Although the landing was cushioned with airbags, Pathfinder decelerated at 40 gravities (Gs).

Pathfinder landed in an area known as Ares Vallis. This site is at the mouth of an ancient outflow channel where potentially a large variety of rocks are within reach of the rover. The rocks would have settled there, being washed down from the highlands, at a time when there were floods on Mars. The Pathfinder craft opened up after landing on Mars (see Fig. 1.1) and released the robotic rover.

The rover on Pathfinder is called Sojourner (see Fig. 1.2). Sojourner is a new class of small robotic explorers, sometimes called microrovers. It is small, with a weight of 22 pounds (lb) [10.5 kilograms (kg)], height of 280 millimeters (mm) (10.9″), length of 630 mm (24.5″), and width of 480 mm (18.7″). The rover has a unique six-wheel (Rocker-Bogie) drive system developed by Jet Propulsion Laboratories (JPL) in the late 1980s. The main power for Sojourner is provided by a solar panel made up of over 200 solar cells. Power output from the solar array is about 16 watts (W). Sojourner began exploring the surface of Mars in July 1997. Previously this robot was known as Rocky IV. The development of this microrover robot went through several stages and prototypes including Rocky I through Rocky IV.

■ **1.1** *Mars Pathfinder. Photo courtesy of NASA*

Both the Pathfinder lander and rover have stereo imaging systems. The rover carries an alpha proton X-ray spectrometer that is used to determine the composition of rocks. The lander made atmospherical and meteorological observations and was the radio relay station to Earth for information and pictures transmitted by the rover.

**Mission objectives**  The Sojourner rover itself was an experiment. Performance data from Sojourner determined that microrover explorers are cost efficient and useful. In addition to the science that has already been discussed, the following tasks were also performed:

☐ Long-range and short-range imaging of the surface of Mars
☐ Analysis of soil mechanics
☐ Tracking Mars dead-reckoning sensor performance

■ **1.2** *Sojourner Rover. Photo courtesy of NASA*

☐ Measuring sinkage in Martian soil

☐ Logging vehicle performance data

☐ Determining the rover's thermal characteristics

☐ Tracking rover imaging sensor performance

☐ Determining UHF link effectiveness

☐ Analysis of material abrasion

☐ Analysis of material adherence

☐ Evaluating the alpha proton X-ray spectrometer

☐ Evaluating the APXS deployment mechanism

☐ Imaging of the lander

☐ Performing damage assessment

Sojourner was controlled (driven) via telepresence by an Earth-based operator. The operator navigated (drove) the rover using images obtained from the rover and lander. Because the time delay between the Earth operator's actions and the rover's response was between 6 and 41 minutes depending on the relative positions of Earth and Mars, Sojourner had onboard intelligence to help prevent accidents, like driving off a cliff.

NASA is continuing development of microrobotic rovers. Small robotic land rovers with intelligence added for onboard navigation, obstacle avoidance, and decision making are planned for future Mars exploration. These robotic systems provide the best value per taxpayer dollar.

The latest microrover currently being planned for the next Mars expedition will again check for life. On August 7, 1996, NASA released a statement that it believed it had found fossilized microscopic life on Mars. This information has renewed interest in searching for life on Mars.

## Industrial robots—going to work

Robots are indispensable in many manufacturing industries. For instance, robot welders are commonly used in automobile manufacturing. Other robots are equipped with spray painters and paint components. The semiconductor industry uses robots to solder (spot weld) microwires to semiconductor chips. Other robots (called "pick and place") insert integrated circuits (ICs) onto printed circuit boards, a process known as "stuffing the board."

These particular robots perform the same repetitive and precise movements day in and day out. This type of work is tedious and boring to a human operator. Following operator boredom comes fatigue, and with operator fatigue, errors. Production errors reduce productivity, which in turn leads directly to higher manufacturing costs. Higher manufacturing costs are passed along to the consumer as higher retail prices. In a competitive market the company that provides high-quality products at the best (lower) price succeeds.

Robots are ideally suited for performing repetitive tasks. Robots are faster and cheaper than human laborers and do not become bored. This is one reason manufactured goods are available at low cost. Robots improve the quality and profit margin (competitiveness) of manufacturing companies.

## Design and prototyping

Some robots are useful for more than repetitive work. Manufacturing companies commonly use computer-aided design (CAD), computer-aided manufacturing (CAM), and computer numerical control (CNC) machines to produce designs, manufacture components, and assemble machines. These technologies allow an engineer to design a component using CAD and quickly manufacture the design of the board using computer-controlled equipment. Computers assist in the entire process from design to production.

## Hazardous duty

Without risking human life or limb, robots can replace humans in some hazardous duty service (see Fig. 1.3). Take for example bomb disposal. Robots are used in many bomb squads across the nation. Typically these robots resemble small armored tanks and are guided remotely by personnel using video cameras (basic telepresence system) attached to the front of the robot. Robotic arms can grab a suspected bomb and place it in an explosion-proof safe box for detonation and/or disposal.

Similar robots can help clean up toxic waste. Robots can work in all types of polluted environments, chemical as well as nuclear. They can work in environments so hazardous that an unprotected human would quickly die. The nuclear industry was the first to develop and use robotic arms for handling radioactive materials. Robotic arms allowed scientists to be located in clean, safe rooms operating controls for the robotic arms located in radioactive rooms.

■ **1.3**   *Hazbot. Photo courtesy of NASA*

## Maintenance

Maintenance robots specially designed to travel through pipes, sewers, air conditioning ducts, and other systems can assist in assessment and repair. A video camera mounted on the robot can transmit video pictures back to an inspecting technician. Where there is damage, the technician can use the robot to facilitate small repairs quickly and efficiently.

## Fire-fighting robots

Better than a home fire extinguisher, how about a home fire-fighting robot? This robot will detect a fire anywhere in the house, travel to the location, and put out the fire.

Fire-fighting robots are so attractive that there is an annual national fire-fighting robot competition open to all robotists. The Fire-Fighting Home Robot Contest is sponsored by Trinity College, the Connecticut Robotics Society, and a number of corporations. Typically a fire-fighting robot becomes active in response to the tone from a home fire alarm. During the competitions, its job is to navigate through a mock house and locate and extinguish the fire.

## Medical robots

Medical robots fall into three general categories. The first category relates to diagnostic testing. In the spring of 1992, Neuromedical Systems, Inc., of Suffern, N.Y., released a product called Papnet. Papnet is a neural network tool that helps cytologists detect cervical cancer quickly and more accurately.

Laboratory analysis of pap smears is a manual task. A technician examines each smear under a microscope looking for a few abnormal cells among a larger population of normal cells. The abnormal cells are an indicator of a cancerous or precancerous condition, but many abnormal cells are missed due to human fatigue and habituation.

Scientists have been trying to automate this checking process for 20 years using computers with standard rule-based programming. This was not a successful approach. The difficulty is that the classic algorithms could not differentiate between the complex visual patterns of normal cells and those of abnormal cells.

Papnet uses an advanced image recognition system and neural network. The network selects 128 of the most abnormal cells found on a pap smear for later review by a cytologist.

The Papnet system is highly successful. It recognizes abnormal cells in 97 percent of the cases. Since the reviewing technician is only looking at 128 cells instead of 200,000 to 500,000 cells on a pap smear, the fatigue factor is greatly reduced. In addition, the time required to review a smear is only one-fifth to one-tenth what it was before. The accuracy improves to a rate of 3 percent false negatives as compared to 30 to 50 percent for manual searches.

The second medical category relates to telepresence surgery. Here a surgeon is able to operate on a patient remotely using a specially developed medical robot. The robot has unique force-feedback sensors that relate to the surgeon the feel of the tissue underneath the robot's instruments. This technology makes it possible for specialists to extend their talent to remote provinces of the world.

The third category relates to virtual reality (VR) and enhanced manipulation. With enhanced manipulation the surgeon operates on a patient through a robot. The robot translates all the surgeon's movements. For instance, let's suppose the surgeon moves his or her hand $1''$; the computer would translate that to travel of $\frac{1}{10}''$ or $\frac{1}{100}''$. The surgeon can now perform delicate and microscopic surgical procedures that were once impossible.

## Nanotechnology

Nanotechnology is the control and manipulation of matter at the atomic and molecular level. It is the ability to create electronic and mechanical components using individual atoms. These tiny (nano) components can be assembled to make machines and equipment the size of bacteria. IBM has already created transistors, wires, gears, and levers out of atoms.

How does one go about manipulating atoms? Two physicists, Gerd Binnig and Heinrich Rohrer, invented the scanning tunneling microscope (STM). The tip of the STM is very sharp and its positioning exact. In 1990 IBM researchers used an STM to move 35 xenon atoms on a nickel crystal to spell the company's name, "IBM." The picture of "IBM" written in atoms made worldwide news and was shown in many magazines and newspapers. This marked the beginning of atomic manipulation. As IBM continues to improve its nanotechnology, nanotechrobotics will find many uses in manufacturing, exploration, and medicine.

### Nanotech medical bots

Nanotechnology can also be used to create small and microscopic robots. Imagine robots so small they can be injected into a patient's bloodstream. The robots travel to the heart and begin removing the fatty deposits, restoring circulation. Or the robots travel to a tumor where they selectively destroy all cancerous cells. What are now considered inoperable conditions may one day be cured through nanotechnology.

Another hope of nanotech medical bots (nanobots) is that they may be able to stop or reverse the aging process in humans. Tiny virus-sized nanobots could enter each cell, resetting the cell clock back to 1. Interesting possibilities.

Keep in mind that nanotechnology is an expanding new robotic field itself. Macroscopic and microscopic robots that will do everything from cleaning your house to materials processing and building are being considered. Everyone expects nanotechnology will be creating new high-quality materials and fabrics at low cost.

## War robots

One of the first applications of robots is war. And if forced into a war, we can use robots to help us win, and win fast. Robots are becoming increasingly more important in modern warfare. Drone aircraft can track enemy movements and keep the enemy under surveillance.

The Israeli military used an unmanned drone in an interesting way. The drone was created to be a large radar target. It was flown into enemy airspace. The enemy switched on its targeting radar, allowing the Israelis to get a fix on the radar position. The radar installation was destroyed, making it safe for fighter jets to follow through.

Smart bombs and cruise missiles are other examples of "smart" weaponry. As much as I appreciate Asimov's Three Laws of Robotics, which principally state that a robot should never intentionally harm a human being, war bots are here to stay.

## Robot wars

There are interesting civilian "robot war" competitions. Competitors build radio-controlled robots that are classified by weight and have them fight in one-on-one battles. Winners advance through standard elimination.

Robot Wars was the first robot war competition. The arena for the competition is 30 by 54 ft of smooth asphalt with 8-ft-high walls to protect spectators. For more information on robot wars, see the Robot Wars website http://www.robotwars.com.

Robot battles have caught on so well that there are a number of robot war competitions and websites to visit. Here are a few:

- [ ] Battlebots     `http://www.battlebots.com/`
- [ ] Robotica     `http://tlc.discovery.com/fansites/robotica/robotica.html`
- [ ] MicroBot Wars     `http://microbw.hypermart.net/`

## Civilian uses for robotic drones

Robotic drones and lighter-than-air aircraft (blimps) developed by the military could be put to civilian use monitoring high-crime neighborhoods and traffic conditions. Because the aircraft do not have any human occupants, they can be made much smaller. I feel robotic blimps will be used more often than robotic aircraft because they will be safer to operate. Aircraft need to be moving in order to maintain lift. An out-of-control drone aircraft can become lethal if it flies into anything. Blimps, on the other hand, are safer because they travel slower and float gracefully through the air. If surveillance aircraft become reliable enough, they could also be used to monitor traffic, warehouses, apartment buildings, and street activity in high-crime areas.

## Domestic

Applications for domestic robots are numerous. We all could use robots that clean windows and floors, report and/or do minor home repairs, cook, clean the upholstery, wash clothes, and change the kitty litter. This raises a debatable point. Should we classify our current labor-saving devices like dishwashers, ovens, washing machines, and clothes dryers as robots or machines? I think that at the point that they autonomously gather the materials needed to perform their functions, like getting food from the refrigerator for cooking or picking up clothes around the house for washing, they will have passed from the machine stage and become robots.

## What we haven't thought of yet—the killer application

It is often said, mostly in regard to software, that to gain popularity you need a "killer application." In the olden days of computers, it was word processing and spread sheets. What will be the killer application for robotics, the one application that will make everyone buy a robot? I don't know the answer to this question. I do know that robots will find many more uses and niches that haven't been thought of today. Many applications will not become apparent until robots are so prevalent in society that the application is discovered by a mixture of availability, imagination, and need.

## More uses

Robotic research and development is moving faster than anyone can follow. The Internet is an excellent tool for finding information.

# 2

# Artificial life and artificial intelligence

THE EVOLUTION OF ROBOTICS LEADS TO TWO FAR-REACH-
ing topics, the creation of artificial intelligence and artificial life.

## Artificial intelligence

People dream of creating a machine with artificial intelligence (AI)
that rivals or surpasses human intelligence. I feel neural networks
are the best technology for developing and generating AI in computer systems. This is in contrast to other computerists who see
expert systems and task-specific rule-based systems (programs) as
potentially more viable.

It is an undeniable fact that rule-based computer operating systems
(DOS, Windows, Linux, etc.) and rule-based software are valuable
and do most (close to all) of the computer labor today. Even so, the
pattern matching and learning capabilities of neural networks are
the most promising approach to realizing the AI dream.

Recently it had been forecasted that large-scale parallel processors using a combination of neural networks and fuzzy logic
could simulate the human brain within 10 years. While this
forecast may be optimistic, progress is being made toward
achieving that goal. Second-generation neural chips are on the
market. Recently two companies (Intel Corp., Santa Clara, CA,
and Nestor Inc., Providence, RI), through joint effort, created a
new neural chip called the Ni1000. The Ni1000 chip, released in
1993, contains 1024 artificial neurons. This integrated circuit

has 3 million transistors and performs 20 billion integer operations per second.

## Evolution of consciousness in artificial intelligence

Consciousness is a manifestation of the brain's internal processes. The generation of consciousness in *Homo sapiens* coincides with the evolution and development of neural structures (the brain) in the biological system. A billion years ago the highest form of life on Earth was a worm. Let's consider the ancestral worm for a moment. Does its rudimentary (neural structure) intelligence create a form of rudimentary consciousness? If so, then it's akin to an intelligence and consciousness that can be created by artificial neural networks running in today's supercomputers (see Fig. 2.1).

In reality, while the processing power of supercomputers approaches that of a worm, this has not yet been accomplished. The reason is that it is too difficult to program a neural network in a supercomputer that would use all the computer's processing power.

The worm is unquestionably alive, but is it self-aware? Is it simply a cohesive jumble of neurons replaying an ancestral record imprinted within its primordial neural structure, making it no more than a functional biological automaton?

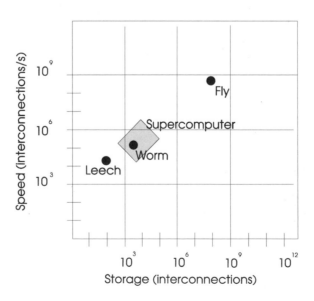

■ **2.1**  *Graph showing supercomputer capabilities*

## Is consciousness life?

This raises a few questions: "Is intelligence conscious?" "Is consciousness life?" It seems safe to say that intelligence has to reach a certain level or critical mass before consciousness is achieved. In any case, artificial neural networks can and will develop consciousness. Whether the time span is 10 years or a 1000 years from now makes no difference; 1000 years is less than a blink of the eye in the evolutionary time line. (Of course, I am hoping for a 10-year cycle so I can see a competent AI machine in my lifetime.) At the point where an artificial neural network becomes conscious and self-aware, should we then consider it to be alive?

# Artificial life

Artificial life (AL) splinters into three ongoing research themes: self-powered neural robots, nanorobotics (may be self-replicating), and programs (software). The most evolved types of artificial life on Earth today are programs. No one has created a self-replicating robot, and nanobots are still years away from implementation. Therefore let's discuss AL programs for the time being.

In AL programs, life exists only as electric impulses that make up the running program inside the computer's memory. Computer scientists have created diverse groups of AL programs that mimic many biological functions (survival, birth, death, growth, movement, feeding, sex) of life. Some programs are called *cellular automations*; others are called *genetic algorithms*.

Cellular automation (CA) programs have been used to accurately model biological organisms and study the spread of communicable diseases like AIDs in the human population. These programs have also been used to study evolution, ant colonies, bee colonies, and a host of other chaos-driven statistics. Chaos algorithms are added into the programs to generate randomness. One interesting application of CA programs is to optimize neural networks running in host computers. It is hoped that these CA programs will one day create and wire large neural network systems in supercomputers.

Genetic algorithms (GAs) evolve in a Darwinian fashion—survival of the fittest. Two compatible GA programs can meet in the computer's running memory, mate, and mix their binary code to produce offspring. If the offspring GA program is as healthy or has greater health than its parents, it will likely survive.

Are these programs alive? It depends upon the definition used for life. What if the programs evolve and develop higher levels of programming? What happens when these programs are encased in and control mobile robots? How about if the robots learn to build copies of themselves (self-replicate)?

### Nanorobotics—are we alive yet?

A *nanobot* is a robot the size of a microbe. IBM is making progress in manipulating atoms and molecules to create simple machines and electronics (transistors and wire). So far, there appears to be no restriction on how small one can make an object. Bacteria-sized robots are theoretically possible.

Some scientists predict silicon life will be the next evolutionary step, replacing carbon life forms on this planet. What we call electronics and robotics will evolve into self-creating, self-replicating silicon life.

Whether or not silicon life becomes the next major evolutionary step on Earth will not be debated here. This chapter will remain focused on the development of artificial intelligence (consciousness) and artificial life.

## A little history

The progression of computer technology over the last five and a half decades is staggering. In 1946 the ENIAC computer filled a large area with electronic equipment. The computer was almost 100 feet (ft) long, 8 ft high, 3 ft deep, and weighed 30 tons. ENIAC contained 18,000 tubes, 70,000 resistors, 10,000 capacitors, 6000 switches, and 1500 electromagnetic relays. ENIAC could perform 5000 additions per second, 357 multiplications per second, and up to 38 divisions per second. Today that same 1946 computer could be condensed on a tiny sliver of silicon less than $1/4''$ square.

Physicist Robert Jastrow stated in *The Enchanted Loom* (New York, Simon & Schuster, 1981) that, "The first generation of computers was a billion times clumsier and less efficient than the human brain. Today, the gap has narrowed a thousand fold."

Science is progressing unrelentingly toward creating AI. Artificial intelligence is something we may see in our lifetime. From the standpoint of creating competent AI, it's a small step to generating superior intelligence in machines. That's a dream, many scientists will tell you, trying to retain the waxing illusion that human intelligence is and forever will be unsurpassed. I don't take any comfort in that illusion. AI is an evolving, uncompromising, unrelenting reality.

## Greater than I

Would we as the human race want to produce an intelligence superior to our own? If you think about it, in the long run we may need to just to survive. Think of the advantages for the first nation that produced an AI machine with an IQ of 300. The AI machine could be given tasks such as improving the national economy, cleaning up the environment, ending pollution, developing military strategy in the event of war, performing medical and scientific research, and, of course, designing still smarter machines than itself. It's possible that the next theory of the universe will not be put forth by a human (as previously done by Albert Einstein) but by a competent AI machine.

## The locked cage

Why is creating a superior intelligence so important? Wouldn't humankind find the answers to all these vexing problems eventually? Perhaps. The necessity of generating a superior AI is best illustrated with a story. I once heard or read this story. I'm afraid I don't remember the author and to him or her I apologize. And if I have changed the story a bit in the retelling, I apologize for that also.

Ten chimpanzees are in a cage. The cage door is locked. To reason how to unlock the lock and open the cage door requires an intelligence quotient (IQ) of approximately 90. Each chimp in the cage has been tested, and each has an IQ of about 60. Could the 10 chimps working together find a way to unlock the cage door? The answer is NO! Intelligence is not accumulative. If it were, the 10 chimps working together would have a combined IQ of 600, more than enough to reason out how to open the cage door. In real life the chimps remain caged.

In the real world we have problems involving global pollution, economics, diseases like cancer and AIDs, the general quest for longevity, and any and all facets of science research that can be substituted for the lock on the cage door. The importance of generating superior AI becomes clearly apparent. The AI may be able to uncover keys to unlock these problems that until then will remain effectively hidden from us.

I don't believe this potential of superior AI is being overlooked by the nations of the world. It's quite possible that the next Manhattan Project undertaken by (hopefully) this country will be for creating superior AI.

We as a race will take no comfort in appearing as intelligent as a chimpanzee to a machine. Science fiction writers have long written on competent AI machines running amok: for instance, the computer HAL in Arthur C. Clarke's *2001, Colossus the Corbin Project,* and the main computer in *Terminator I* and *Terminator II.* So to all the future AI programmers out there reading this book, I have an important message: "Don't forget that off switch!"

## Biotechnology

Advances in biotechnology will soon allow us to alter our own genetics. With this power it becomes possible to enhance our brain to increase our own intelligence. While possible, it opens up the human race to unforeseeable repercussions of gene altering in subsequent generations that may be catastrophic. This makes generating superior intelligence in machines much safer, at least for the time being.

## Neural networks—hype versus reality

Neural networks have been overhyped since their inception. So it's easy to dismiss my remarks concerning AI, AL, and neural networks as just more of the same, as people have been doing for years. And it is true that people have predicted the emergence of humanlike intelligence in machines.

If progress continues as rapidly as it has in the past 50 years, I believe human levels of intelligence in machines will be achieved within 50 years.

## What are neural networks?

I have discussed neural networks without defining them. Here is their definition. *Neural networks* are artificial systems (hardware and software) that function and learn based upon models derived from studying the biological systems of the human brain. Networks may be implemented in either software/operating systems or hardware. In mimicking the biological systems of the brain, neural networks have taken strides in building the sensory foundations needed for AI, such as machine vision, voice recognition, and speech.

Neural networks can be trained to perform visual recognition. They can learn to read or perform quality control by visual analysis of parts. One such example is Papnet, discussed in Chap. 1. Other networks can be taught to respond to verbal commands (speech recognition) and generate speech. Statistical nets can predict the future behavior or probability of complex nonlinear systems based on historical examples. These networks have been used to predict oil prices, monitor aircraft electronics, and forecast the weather. Networks have also successfully been employed to evaluate the stock market, mortgage loan applicants, and life insurance contracts better than standard rule-based expert-system programs.

## What is artificial intelligence?

This is a legitimate question. We most certainly will develop neural networks that are intelligent before we develop nets that become conscious. So in attempting to create neural networks that are intelligent or demonstrate intelligence, what criteria should one use to determine if this goal has been achieved?

Alan Turing, a British mathematician, devised an interesting procedural test that is generally accepted as a valid way to determine if a machine has intelligence. The test is conducted as follows: A person and the machine hold a conversation by typing messages to one another via a teletype. If the machine can carry on a conversation without the person being able to determine whether a machine or person exists at the other teletype, the machine can be classified as intelligent. This is called the Turing test and is one criteria used to determine AI.

Although the Turing test is well accepted, it isn't a definitive test for AI. There are a number of "completely dumb" language processing programs that come close to passing the Turing test. The most famous program is named ELIZA, developed by Joseph Weizenbaum at the Massachusetts Institute of Technology (MIT). ELIZA simulates a psychologist, and you are able to conduct a conversation with ELIZA. For instance, if you typed to ELIZA that you missed your father, ELIZA might respond with "Why do you miss your father?" or "Tell me more about your father." These responses may lead you to believe that ELIZA understands what you have said. It doesn't. The responses are clever programming tricks constructed from your statements.

Therefore, if we like, we could do away with the Turing test and consider a different criterion. Perhaps consciousness or self-awareness would be a better signpost of intelligence. A self-aware machine would certainly know that it is intelligent. Another criterion, more simple and direct, and the one that is used in this book, is the ability to learn from experience.

Of course, we could abandon logical approximations entirely and state that intelligence is achieved in systems that develop a sense of humor. As far as I know humans are the only animals that laugh. Perhaps humor and emotion will end up being the truest test of all.

## Using neural networks in robots

So how do neural networks help our robotics work today? Well, we're a way off from creating competent AI, let alone putting it into one of our robots. But neural technology can control robotic function, and, in many cases, can perform superiorly to standard central processing unit (CPU) control and programming. By using neural networks in our robots, we can have our robots perform small operational miracles without the use of a standard computer, CPU, or programming. In Chap. 6 we will design a two-neuron fuzzy logic system that can track a light source. Place this system on a mobile robot, and the robot will follow a light source anywhere. Also in Chap. 6 we discuss BEAM robotics and Mark Tilden, who designs transistor networks (nervous networks) that allow legged robots to walk and perform other functions. Another neural process that is making great strides is called *subsumption architecture,* which uses layered stimulus response.

## Tiny nets

Small neural network programs can also be written in microcontrollers. For more information on these microcontrollers see Chap. 6.

## Neural-behavior-based architecture

Behavior-based architecture, developed by Walter Grey, illustrates that relatively simple stimulus-response neural systems when placed in robotics can develop high-level, complex behaviors. Subsumption architecture, an offshoot of behavior-based architecture developed by Dr. Rodney Brooks at MIT, is also covered more fully in Chaps. 6 and 8.

# 3

# Power

ROBOTS NEED POWER TO FUNCTION. MOST ROBOTS USE electric power. The two main sources of untethered electric power for mobile robotics are batteries and photovoltaic cells. In a few years fuel cells will become a third electric power source for robotics.

## Photovoltaic cells

Photovoltaic cells, commonly called solar cells, produce electric power from sunlight. A typical solar cell produces only a small amount of power, a few milliamperes at a potential difference of about 0.7 volts (V). Solar panels (modules) use many solar cells strung together to produce an appreciable power. The same is true with robotics. If enough solar cells are strung together, in series and parallel, sufficient power can be generated to operate a robot directly.

Solar-powered robots need to be designed as small as possible while still being able to perform their designated functions. They should be constructed using high-strength, lightweight materials and low-power electronics.

The greater the weight reduction and the smaller the electric power consumption needed for operation and locomotion, the more viable the use of solar energy becomes. However, reduced weight and power consumption are important in the design of every robot. Lightweight, low-power robots are able to operate longer on a given power supply than their heavier, power-hungry counterparts.

Solar cells can also indirectly power a robot by being used as a power source for recharging the robot's batteries. This hybrid power supply reduces the required capacity of the solar cells needed to operate the robot directly. However, the robot can only function for a percentage of the time that it spends recharging its power supply.

We can also utilize solar cells by combining the technologies of direct and indirect power. Here we build what is commonly called a *solar engine.* The circuit is simple in function. The main components are a solar cell, main capacitor, and a triggering circuit. The solar cell when exposed to light begins charging a large capacitor. The solar cell/capacitor provides electric power to the rest of the circuit. As the charge on the capacitor increases, the voltage to the circuit also rises until it reaches a preset level that triggers the circuit. Once the circuit is triggered, the power stored in the capacitor is dumped through the main load. The cycle then repeats. The solar engine may be used in a variety of innovative robotic designs.

## Building a solar engine

The solar engine is commonly used as an onboard power plant for BEAM-type robots, sometimes called living robots (see the discussion about BEAM robots in Chap. 8). The inspiration for this solar engine originated from Mark Tilden, who originally designed a solar engine. Another innovator was Dave Hrynkiw from Canada, who modified the solar engine design to power a solar ball robot. I liked the electrical function so much that I decided to design my own solar engine. In doing so I was able to create a new circuit that improved the efficiency of the original design.

Figure 3.1 is the schematic for the solar engine. Here is how it works. The solar cell charges the main 4700-microfarad ($\mu$F) capacitor. As the capacitor charges, the voltage level of the circuit increases. The unijunction transistor (UJT) begins oscillating and sending a trigger pulse to the silicon controlled rectifier (SCR). When the circuit voltage has risen to about 3 V from the main capacitor, the trigger pulse is sufficient to turn on the SCR. When the SCR turns on, all the stored power in the main capacitor is dumped through the high-efficiency (HE) motor. The motor spins momentarily as the capacitor discharges and then stops. The cycle repeats.

The solar engine circuit is simple and noncritical. It may be constructed using point-to-point wiring on a prototyping breadboard. A printed circuit board (PCB) pattern is shown in Fig. 3.2 for those who want to make the PCB. The solar engine kit (see parts

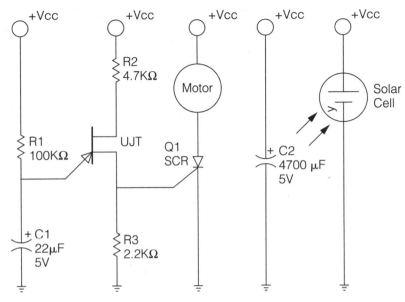

■ **3.1** *Schematic for solar engine*

list) has a PCB included. Figure 3.3 shows the PCB parts placement. The complete solar engine is shown in Fig. 3.4.

### Parts list for solar engine

- ☐ (1) 2N2646 UJT transistor
- ☐ (1) 2N5060 SCR
- ☐ (1) 22-µF cap
- ☐ (1) 0.33-F cap
- ☐ (1) DC motor
- ☐ (2) Solar cells
- ☐ (1) PCB
- ☐ (1) R1 200K ohm, ¼ watt (W)
- ☐ (1) R2 15K ohm, ¼ W
- ☐ (1) R3 2.2K ohm, ¼ W

### High-efficiency motor

Not all electric motors are HE. For instance, the small electric motors sold at your local Radio Shack are of the low-efficiency type. There is a simple way to determine if a motor is an HE type. Spin the rotor of the motor. If it spins smoothly and continues to spin momentarily when it is released, it is probably an HE type.

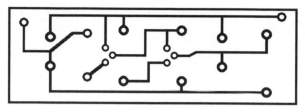

■ **3.2** *PCB foil pattern*

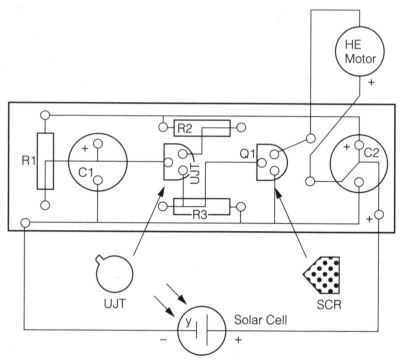

■ **3.3** *Parts placement on PCB*

If when you spin the rotor it feels clunky or there is resistance, it probably is a low-efficiency type.

### Caveats regarding the solar engine

The solar cell used in this circuit is high voltage, high efficiency. Typically solar cells supply approximately 0.5 to 0.7 V at various currents depending upon the size of the cell. The solar cell used in

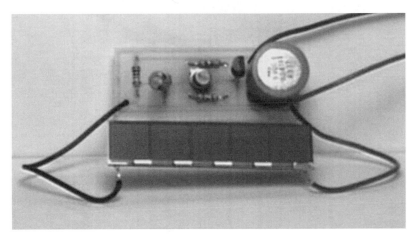

■ **3.4** *Complete solar engine*

this circuit is rated at 2.5V, but I've seen it charge a capacitor up to 4.3V under no-load conditions.

I'm sure some people planning to build this circuit are already thinking about adding a few more solar cells to speed charging. You should not do this. Adding solar cells will increase the current and will speed charging for the first cycle only. In order for the circuit to recycle, the current through the SCR must stop (or at least be very minimal) for the SCR to close. If there is too much current being supplied by the solar cell(s), the SCR will stay in the on condition. If this happens, the electric energy from the solar cell will continually flow through the SCR and be dissipated. Electric energy will not build on the capacitor, and the solar engine circuit will stop cycling.

The components used in the circuit are balanced for proper operation. One component you may change is the main capacitor. You may use smaller values for quicker charge-and-discharge cycles. A larger capacitor (or capacitor bank) will store more electric power and perform more work, but be aware that when using larger-value capacitors it will take that much longer to go through the charge-discharge cycle.

### Uses

A solar engine circuit may be used in many novel and innovative ways: as an onboard power plant for a solar racer, to supply power to a relay station, as a flashing light-emitting diode (LED) buoy, as a motor for robotic locomotion, or as the demo circuit pictured here does (see Fig. 3.5) to spin an American flag.

The attractiveness of the solar engine circuit is that it operates perpetually, or at least until one of the components breaks, which means it should operate for years.

# Batteries

Batteries are by far the most commonly used electric power supply for robotics. Batteries are so commonplace that it's easy to take them for granted. An understanding of batteries will help you choose batteries that will optimize your robot's design. The rest of this chapter will examine batteries.

There are hundreds of different kinds of batteries. We will look at the most common batteries employed for hobbyist use: carbon-zinc, alkaline, nickel-cadmium, lead-acid, and lithium.

## Battery power

Regardless of battery type, battery power is measured in amp-hours, that is, the current (measured in amps or milliamps) multiplied by the time (hours) that current is flowing from the battery. What does that mean to us? Well, it's pretty straightforward. Suppose a battery is rated at 2 amp-hours (Ah). This means the battery can supply 2 A of current for 1 h. If we reduce the current draw from the battery to only 1 A, the battery will then last 2 h. If the current is further reduced to 500 milliamps (mA), the battery will last 4 h. If you do the math for the three different scenarios, you will see battery life (time) is in direct proportion to the current draw:

| Current | $\times$ | time | = | battery rating |
|---------|----------|------|---|----------------|
| 2 A | | 1 h | | 2 Ah |
| 1 A | | 2 h | | 2 Ah |
| 0.5 A | | 4 h | | 2 Ah |

It becomes an easy matter to rearrange the equation to tell you how long a battery will last given a particular current draw. For instance, suppose your robot draws 0.35 A (350 mA). If you are using the same battery (2 Ah) as just discussed, then divide the battery rating (2 Ah) by the current draw (0.35 A) to find the time the battery will last (5.7 h). Mileage will vary. Batteries provide more access to their electric power rating when used intermittently, which allows time for the battery to recuperate. Continuous duty is efficient if the load

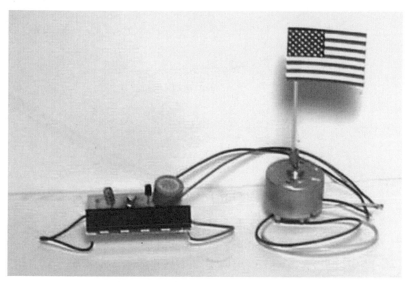
■ **3.5** *Solar engine flag spinner*

is light. For robotics, especially when powering drive motors and components, we often don't have this option. In this case one tries to provide greater battery capacity.

## Battery voltage

Battery voltage varies throughout the life of a battery. If you measure the voltage on a fresh D-sized 1.5V alkaline battery, it will read approximately 1.65 V. As the battery discharges, its voltage drops. The battery is considered "dead" when the voltage drops to 1.0 V. Typical discharge curves for carbon-zinc, alkaline, and nickel-cadmium (NiCd) batteries are illustrated in Fig. 3.6.

Notice that a fresh NiCd 1.5V battery actually delivers about 1.35 V. While its initial voltage is lower, its discharge curve is fairly flat compared to that of carbon-zinc and alkaline batteries delivering a constant 1.2 V.

## Primary batteries

Primary batteries are one-time-use batteries. The batteries we will look at in this class deliver 1.5 V per cell. They are designed to deliver their rated electrical capacity and then be discarded. When building robotic systems, discarding depleted primary batteries can become expensive. However, one advantage to using primary batteries is that

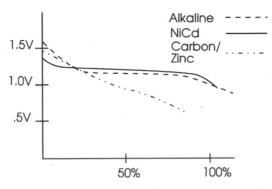

■ **3.6** *Discharge curves for common batteries*

they typically have a greater electrical capacity than rechargeables. If one is engaged in a function (i.e., a robotic war) that requires the highest power density available for one-shot use, primary batteries may be the way to go.

### Rating primary batteries

As you may have guessed, there are a number of primary batteries available. The differences in batteries relate to the chemistry used in the battery to produce electricity. The choice of a primary battery is a tradeoff of price versus energy density, shelf life, temperature range, discharge slope, and peak current capacity.

**Carbon-zinc**  At the low end of primary batteries is the carbon-zinc battery. This battery hasn't changed much since 1868 when it was developed by George Leclanche. The carbon battery has a low energy density [1 to 2 watthours per cubic inch ($Wh/in^3$)], poor high-current performance, sloping discharge curve, and bad low-temperature performance. It is inexpensive but obsolete.

**Alkaline-manganese**  This is simply an alkaline battery. It has an energy density of 2 to 3 $Wh/in^3$, improved low-temperature performance, and a sloping discharge curve, not as severe as carbon-zinc batteries. Its cost is moderate.

**Lithium**  This is a premium battery with a high energy density (8 $Wh/in^3$), excellent low-temperature and high-temperature performance, and a long shelf life (15 years). It is also lightweight, but expensive.

## Secondary batteries

Secondary batteries are rechargeable. The most common rechargeable batteries are NiCds and lead-acid. We will start with NiCd batteries.

One disadvantage to NiCd batteries is that they have a lower voltage, 1.2 V per cell. So a C cell battery will deliver about 1.2 V instead of 1.5 V. The effect becomes more pronounced when using multiple cells. For instance, a "9V" NiCd battery made from six NiCd batteries will deliver approximately 7.2 V.

Automotive lead-acid batteries are rechargeable but are not suitable for robotics. The reason is that automotive batteries are not designed to be completely discharged (run down) before being recharged. These batteries can supply high currents for short periods of time (car starting) and need to be recharged almost immediately.

Completely discharging the electric power a rechargeable battery contains before recharging the battery is called *deep cycle*. There are deep-cycle lead-acid batteries available, mostly because of the solar power industry, but you will find these batteries carry a higher price tag. When building robotic systems, you should use deep-cycle rechargeable batteries.

Secondary batteries, while initially more expensive, are cheaper in the long run. Typically secondary batteries can be recharged 200 to 1000 times. In many cases a simple recharging circuit can be built into the robot so that it becomes unnecessary to remove batteries for charging.

### Rating secondary batteries

**NiCd**   NiCd batteries and sealed lead-acid batteries are the most common rechargeables, with NiCd batteries being more popular. Both types of batteries have lower energy densities than primary batteries.

NiCd batteries only provide 1.2V per cell, in comparison to primary batteries which provide 1.5V per cell. Manufacturers claim that NiCd batteries are good for 200 to 1000 charge-recharge cycles. However, NiCd batteries will die fast if they aren't recharged properly. The life expectancy of NiCd batteries is 2 to 4 years. Without use, a fully charged NiCd battery will lose its charge in 30 to 60 days.

NiCd batteries are designed to be recharged at 10 percent of their rated capacity. This means that if a particular NiCd battery is rated at 1 Ah, it is safe to recharge the battery at 100 mA (1 A/10 = 100 mA). The terminology used to describe the above recommended recharge rate is "C/10."

NiCd batteries are designed to be charged using a constant current at the C/10 rates. Because of inefficiencies, it is necessary to charge the battery for 14 h to get a full charge. While manufacturers claim that it is OK to overcharge NiCd batteries at the C/10 rate, most engineers recommend switching over to a trickle charge after the initial 14 h at C/10. A trickle charge is usually rated at C/30, or 1/30 of the battery's capacity. A trickle charge for our 1-Ah battery would be around 33 mA (1 A/30 = 33.3 mA).

**Memory effect**   A disadvantage to NiCd batteries is the memory effect. If one repeatedly recharges a NiCd battery before it has completely discharged, the battery forms a memory at that recharge level. It then becomes difficult to discharge the battery past that remembered level. Obviously this can severely limit the battery's capacity. To correct that problem the battery must be completely discharged, by leaving a load connected to the battery for several hours. Once the battery is complete discharged, it can be charged normally and will function properly.

**Lead-acid**   Gelled-electrolyte battery cells (gel-cells) are similar to automotive batteries. They are sealed, maintenance-free, lead-acid batteries. They don't make gel-cells in the familiar D, C, AA, AAA, or 9V battery cases. Gel-cells are typically larger and may be used in larger robots.

Gel-cells are available in numerous voltage ratings, from 2V to 24V, and current capacities. These batteries may be charged with a current-limited constant voltage or a constant current like NiCd batteries. Typically to charge a gel-cell, one applies a fixed 2.3V to 2.6V per cell. Initially the battery will draw a high current that tapers down as it charges. When fully charged, the battery need only draw a trickle charge (approximately C/500) to maintain itself in a fully charged state.

Gel-cell batteries vary from manufacturer to manufacturer. To safely recharge a gel-cell, you should check the manufacturer's recommendation. In general, a simple charging device can be made using an LM317 voltage regulator. A fixed voltage (2.3V per cell), constant current C/10 is applied to the battery. When the battery reaches a full charge, the constant current source is removed and a regulated voltage is applied.

Many gel-cell batteries do not like to be deep cycled. Therefore it becomes necessary to monitor battery voltage under load. When the battery voltage drops by a specified amount (check the manufacturer's data sheet), it needs to be charged.

### In general

Most robotists use alkaline batteries when primary batteries are called for and NiCd batteries when secondary batteries are needed.

### Building a NiCd battery charger

NiCd battery chargers are inexpensive. Typically it is not worth the time and effort to build a stand-alone charger for common-size batteries such as AAA, AA, C, D, and 9V. However, if one wishes to incorporate a built-in charger for a robot, then knowing how to build a custom battery charger is important. While most inexpensive chargers will charge batteries only at the C/10 rate, even after the batteries have received a full charge (14 h), the charger we will build will drop the current down to a C/30 rate after the batteries are fully charged. This is the recommended procedure for charging NiCd batteries. This will help ensure a long service life to your rechargeable battery.

The following information will allow you to design a system for charging a custom NiCd battery pack.

The prototype charger shown in Fig. 3.7 is a stand-alone unit for illustration purposes. The design can easily be placed inside a robot. The robot will need to have a power socket that connects to the power supply. In between the socket and power supply, you should add a double-pole double-throw (DPDT) switch. The DPDT switch connects the power supply to either the robot's circuitry or the charger. This prevents powering the robot, which would reduce the current flow to the batteries, while the batteries are being charged (see Fig. 3.7).

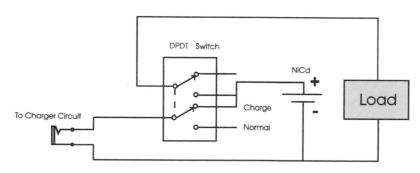

■ **3.7** *DPDT switch controlling charging to battery pack*

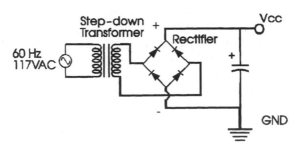

**■ 3.8** *Basic power supply for charger circuit*

The power for the charger may be supplied by either a standard transformer or a VDC plug-in wall transformer. I would choose a wall transformer because it supplies a DC voltage. If you are using a standard transformer, you must build the power supply, using a line cord, switch, fuse, bridge rectifier, and smoothing capacitor.

In either case you should match the transformer (or wall transformer) power output to the battery pack you are charging. Matching the voltage and current to the battery pack reduces the power the LM317 must dissipate; for example, you wouldn't want to use a 12V transformer to charge a 6V battery pack.

Figure 3.8 is a basic VDC power supply for the charger. The power supply can be made to provide either 6V, 12V, 18V, 24V, or 36V depending upon the transformer, bridge rectifier, and capacitor chosen.

The charger circuit is illustrated in Fig. 3.9. It uses an LM317 voltage regulator and a current-limiting resistor. The resistance needed to be provided by the current-limiting resistor depends upon the current needed to charge the battery.

### Current-limiting resistor

Most NiCd battery manufacturers recommend charging the battery at 1/10 of its rated capacity, referred to as C/10. So if an AA battery is rated at 0.850 Ah, it should be charged at 1/10 that capacity, or 85 mA, for 14 h. After the batteries are fully charged, manufacturers recommend dropping the current to around C/30 (1/30 of battery capacity) to keep them fully charged without overcharging or damaging the batteries in any way.

For our example, we will configure the charger to recharge four C cells in series. Each C cell is rated at 2000 mA. So our C/10 rate is 200 mA. The typical voltage rating from this battery is approxi-

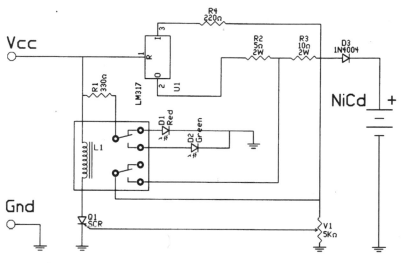

■ **3.9** *Schematic of charger circuit*

mately 1.3V (4 × 1.3V = 5.2V). We can use a 6V transformer with at least a 200-mA output.

To calculate the resistance to be provided by the current-limiting resistor, use the formula

$$R = 1.25/Icc$$

where Icc is the desired current. Plugging in our 200 mA (0.2 A) yields

$$1.25/0.2 = 6.25 \text{ ohms}$$

The resistance of the current-limiting resistor for this charger should be around 6.25 ohms. In the schematic (Fig. 3.9), this resistor is labeled R2. Notice the R2 value listed in the schematic is 5 ohms. You should choose a common resistor value as close as possible to the calculated value.

### C/30 resistor

To drop the current to a C/30 range, we add another resistor whose value is 2R, or about 12.5 ohms. In the schematic this resistor is labeled R3. Again a resistor with the closest value to the calculated value is used. In this case the value is 10 ohms.

### How the charger works

The charger uses an LM317 voltage regulator as a constant current source. The C/10 current-limiting resistor is identified as R2 in the schematic (see Fig. 3.9). R2 you will notice is only 5 ohms as compared to the calculated 6.25 ohms. This standard value is close enough to the calculated value for proper operation. The C/30 resistor is R3 on the schematic. Again the standard value of 10 ohms is close enough to the calculated value for proper operation. Later on we will see that it's possible to fast-charge the batteries because of the voltage-sensing capacity of the circuit.

V1 is a 5K-ohm potentiometer. It is set to trigger the SCR when the NiCd batteries are fully charged. The SCR, once triggered, allows current to flow through a DPDT relay.

When power is applied to the circuit, current flows through the LM317 charging the batteries at a C/10 rate. Resistor R3 is shorted by one-half of the DPDT relay. Current also flows through resistor R1, which is a current-limiting resistor for light-emitting diodes (LEDs) D1 and D2. Upon power-up, the red LED D1 will be lit. The red LED indicates that the circuit is charging.

As the batteries charge, the voltage drop across V1 becomes greater. After about 14 h, the voltage drop across V1 is great enough to trigger the SCR. When the SCR is triggered, current flows through the coil of the DPDT relay. The relay switches, causing the red LED to go out and the green LED to turn on. The green LED signals that the batteries are fully charged. The other half of the relay switches, opening up the short on resistor R3. With R3 now in the current path, the current flowing to the NiCd batteries is cut to a C/30 level. Diode D3 prevents any current from the batteries flowing back into the circuit.

### Determining the trigger voltage from V1

For the circuit to function properly, the SCR must turn on when the NiCd batteries are fully charged. The easiest (best) way to do this is to place depleted batteries in the charger, charge the batteries for 14 h, and then adjust V1. When the batteries are fully charged, slowly turn V1 until the relay clicks and the green LED turns on.

### Design notes

When building a charger for your application, keep these points in mind. The main considerations are choosing the C/10 and C/30 current-limiting resistors. Use the given formulas for selecting these values. Current-limiting resistors should be rated around 2 W.

If the charging current is high (greater than 250 mA), heat-sink the LM317. If the charger is switched on without the NiCd batteries being connected, the relay will switch immediately, turning on the green LED and providing a C/30 current.

When building a charger for higher voltages, increase the value of R1 proportionally to limit the current flowing through the LEDs. For instance, for a 12V unit make R1 680 ohms; for a 24V unit make R1 1.2K ohms.

At high voltages you may need a low-ohm-value, current-limiting resistor connected to the DPDT relay. Measure the C/10 and C/30 current flowing to the batteries. These measurements will ensure that the proper current is being supplied to the batteries.

### Series and parallel charging

How the batteries are configured determines the voltage and current of the transformer one should use. If you have eight C battery cells in parallel, you need to multiply the current requirements of each individual cell by 8. If the cell is rated at 1200 mAh, the C/10 requirement per battery is 120 mA. For eight cells in parallel, you need close to 1 A ($8 \times 120$ mA = 960 mA = 0.96 A) of current. The voltage required is just 1.5 V. The ideal transformer's output would be 1.5V at 1 A. If the eight cells were held in series, the current requirements would be 120 mA at 12V.

### Fast charger

Many of today's NiCd batteries are capable of accepting a fast charge provided that the circuit can sense when the batteries are fully charged and drop the current to C/30. Typically to fast-charge a battery, you double the current for half the time. So you charge a battery at C/5 for 7 h.

Although I haven't tried the above circuit for fast charging, there is no reason why it shouldn't work. You may want to start with a C/10 charging current and adjust V1, and then switch resistor R2 for a resistor with half the value.

### Parts list

☐ U1 LM317 voltage regulator

☐ L1 DPDT relay (5V or 12V)

☐ D1 Red LED

☐ D2 Green LED

☐ D3 1N4004

☐ Q1 SCR

☐ V1 5K-ohm PC-mounted potentiometer

☐ R1 330 ohms, ¼ W

☐ R2 5 ohms, 2 W

☐ R3 10 ohms, 2 W

☐ R4 220 ohms, ¼ W

☐ Wall transformer (see earlier in this chapter)

### Building a solar-powered battery charger

Once you have designed a battery charger for a rechargeable battery pack, you can convert it to a solar-powered battery charger. You need to replace the step-down transformer (or wall transformer) with a combination of solar cells that will equal the power delivered by the transformer. Points to keep in mind when planning a solar power system are

1. The average illumination received by the solar panel
2. Hours of illumination needed to recharge power supply versus work period

## Fuel cells—batteries with a fuel tank

Fuel cells and batteries are both electrochemical devices that convert chemical energy into electric energy. In the battery, the chemical reactants are stored internally. When the reactants are exhausted, the battery is replaced (or in some cases recharged). Fuel cells use reactants (fuel) that are stored externally. As long as fuel is supplied to a fuel cell, it will (in theory at least) continue to generate electricity.

When a fuel cell starts running low, it can be simply refilled with fuel, much like today's automobile. A fuel cell–powered robot will be able to get back to work quickly as compared to another robot that will be out of service while its batteries recharge.

Figure 3.10 is a schematic of an alkali fuel cell. This is the type of fuel cell used in U.S. spacecraft. The first thing you may notice is that the anode is labeled $(-)$ and the cathode is labeled $(+)$. When I first started looking at fuel cell schematics, I found this confusing. Actually I though it was a mistake, but after looking at a few dozen schematics with the same error, I realized it couldn't be a mistake. Accordingly, I checked the definition of cathode in the *Oxford Dictionary of Current English*. It reads: "Cathode: 1. Negative electrode in an electrolytic cell. 2. Positive terminal of a battery." I only bring this up so that you don't become confused when studying other schematics of fuel cells, since all the schematics I've seen so far follow this convention.

Fuels cells have many applications. Just about anything that uses batteries will benefit from fuel cell technology. A few applications already in the works are an aluminum/air fuel cell to be used for cellular phones and a fuel cell for laptop computers. Fuel cells offer extended run times and improved performance.

## If not now, when?

With all the wonderful attributes of fuel cells, where are they? Why don't we find fuel cells in our laptop computers, video cameras, and cell phones? While fuel cell technology has improved over the last decade, fuel cells still are not competitive (read cost-effective) with existing technologies.

One of the more advanced fuel cell designs uses a proton exchange membrane (PEM) electrode material from DuPont called Nafion. The PEM raw material costs approximately $100 per square foot. Reducing

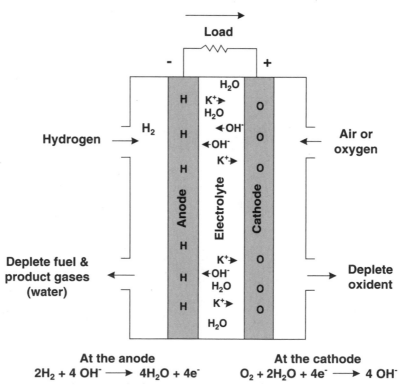

**At the anode**

$$2H_2 + 4\ OH^- \longrightarrow 4H_2O + 4e^-$$

**At the cathode**

$$O_2 + 2H_2O + 4e^- \longrightarrow 4\ OH^-$$

■ **3.10** *Alkali electrolyte [potassium hydroxide (KOH)] fuel cell* existing membrane cost and developing other PEM materials are on a high-priority list for creating competitive fuel cells.

Platinum is an expensive metal. Fuel cell electrodes are typically coated or plated with platinum. The platinum coating is a catalyst that facilitates the electrochemical processes in the fuel cell.

Where we will start seeing fuel cell technology is in the automotive industry. All major automobile companies have ongoing research and development for implementing fuel cell technology. The list of industrial companies working on fuel cell technology reads like a who's who of science research.

Expect to see fuel cell automobiles hitting the market in 2003. Ballard Power Systems of Canada, a major player in PEM fuel cell technology, has been running a fleet of fuel cell buses. In 2003, 55 fuel cell buses are contracted to arrive in California. Ballard has partnered its technology with other companies like DaimlerChrysler and Ford Motor company to produce fuel cells for automotive use. Ballard recently opened a plant that is scheduled to produce 160,000 commercial fuel cells annually.

Honda has scheduled a production fuel cell car for 2003. It will use an existing electric drive system developed for battery-powered cars and retrofit it with fuel cell technology.

There is much enthusiasm and support for continued research and development of fuel cells. Before leaving office President Clinton, along with Congress, allocated $100 million for fiscal year 2001 for the continued development of fuel cell technology.

As fuel cell technology filters down into work-a-day applications, like cell phones, video cameras, and laptop computers, we can then start putting them to work in our robots.

# 4

# Movement and drive systems

IN THIS CHAPTER WE WILL LOOK AT A FEW MOVEMENT AND drive components that may be used in robots. All the components discussed in this chapter either have sample circuits contained in this chapter or are used in robots elsewhere in this book. Here is a list of the components we will work with: air muscles, nitinol wire, stepper motors, geared direct current (DC) motors, servo motors, and solenoids.

## Air muscles

An *air muscle* is a simple pneumatic device developed in the 1950s by J. L. McKibben. Like biological muscles, air muscles contract when activated. An interesting fact about air muscles is that they provide a reasonable working copy of biological muscles, so much so that researchers can use air muscles attached to a skeleton at primary biological muscle locations to study biomechanics and low-level neural properties of biological muscles. In published papers, air muscles are also referred to as McKibben air muscles, McKibben pneumatic artificial muscles, and Rubbertuator. I will refer to them simply as air muscles.

### Applications

Air muscles have applications in robotics, biomechanics, artificial limb replacement, and industry. The principal reasons why experimenters and hobbyists will like air muscles are their ease of use (as compared to standard pneumatic cylinders) and their simple construction. Air muscles are soft, lightweight, and compliant; they have a high power-to-weight ratio (400:1); and they can be twisted axially and used on unaligned mountings and provide contractive force around bends.

## How air muscles work

The two primary components to the air muscle are a soft, stretchable inner rubber tube and a braided polyester mesh sleeve (see Fig. 4.1). The rubber tube is called an "internal bladder" and is positioned inside the braided mesh sleeve.

Additional components of the air muscle are an air fitting located on one end and two mechanical fittings (loops), each located on one end of the air muscle, that allow one to attach the air muscle to devices.

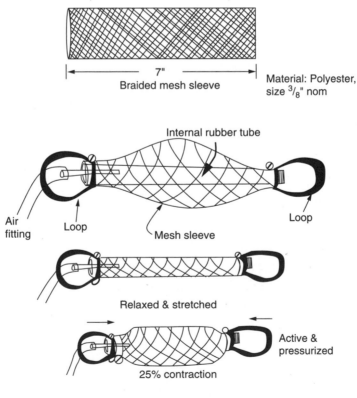

■ **4.1** *Air muscle construction and function*

When the internal bladder is pressurized, it expands and pushes against the inside of the braided mesh sleeve, forcing the diameter of the braided mesh to expand. The physical characteristic of the mesh sleeve is that it contracts in proportion to the degree its diameter is forced to increase. This produces the contractive force of the air muscle.

It is important to note that to operate properly, the air muscle must be in a stretched or loaded position when it is in a resting state. If not, when the air muscle is activated, there will be little if any contraction. Typically the air muscle can contract to approximately 25 percent of its length.

## Nitinol wire

Nitinol is a metal that belongs to a class of materials called shaped memory alloys (SMAs). Nitinol is commonly sold in wire form. When heated, the material can contract up to 10 percent of its length. The contraction of the material produces linear motion. In addition to the contraction property, the material also exhibits a shaped memory effect (SME).

The SME is a unique property of this alloy. When heated to its critical transition temperature, the material automatically returns to a predefined shape. The predefined shape is one that the material is trained (heat annealed) to remember. The material is formed into the training shape. The material is then forcibly confined to the training shape as the material is annealed (heated) above its transitional temperature. This realigns the crystalline structure to the shape. Now the object will return to this shape whenever it is heated to its transition temperature. So a trained object could be twisted and folded out of shape and then heated to return the object back to its original shape.

These unique properties of SMAs rely upon the crystalline structure of the material. The shape-resuming force approaches 22,000 pounds per square inch (lb/in$^2$). It's very unlikely that anyone will be working with such large cross sections of material. Even thin wires of the material produce an impressive force. For instance, a 6-mil wire generates a contractive force of 11 ounces.

When nitinol wire contracts up to 10 percent of its overall length, its volume remains constant. As the wire contracts, its diameter increases proportionally, keeping the net volume of the wire constant.

The easiest way to heat nitinol wire is by passing an electric DC current through it (see Fig. 4.2). However, using a steady DC current

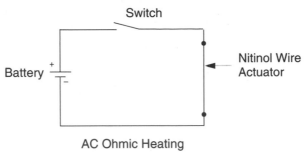

Switch

Battery

Nitinol Wire Actuator

AC Ohmic Heating

■ **4.2** *DC power to nitinol wire*

for an extended period of time can damage the wire due to uneven ohmic heating. Proportional control and steady-state contraction of the material (without damage) can be achieved using a pulse-width modulation (PWM) circuit to supply the electric current.

Some robotists have used nitinol wire to create a motorless hexapod walking robot. While the robot can walk, it does so exceedingly slowly due to the time required for cycling (heating and cooling) of the nitinol material. The hexapod walker robot is truly a flyweight (a few ounces at most) robot, neither structurally strong nor powerful enough to carry its own power supply.

While nitinol may be impractical for use in a hexapod walker, it is appropriate for many other robotic applications. To learn more about the capabilities of this remarkable material, let's look at a few commercial products that utilize the contractile ability of the material. Figure 4.3 shows a mechanical butterfly. Nitinol wire adds movement to the wings. The butterfly may be connected to a solar engine (see Chap. 3) for power to create an interesting robotic application.

Figure 4.4 shows a rocker ball demonstration device. The nitinol actuator operates about 20,000 cycles per day and will last for years.

Nitinol wire loops can be used to produce rotary motion. Figure 4.5 illustrates a simple heat engine. The nitinol loop is guided by a groove in each wheel. The smaller wheel is made of brass for good heat conduction. When the smaller wheel is placed in hot water, the wheels begin to spin. The heat engine can also function using solar energy. Focusing sunlight from a 3″ magnifying glass onto the brass wheel will also activate the engine.

Nitinol can also be used to physically close mechanical push-button switches, as an actuator in production of lightweight air valves, and in many other linear-motion applications.

■ **4.3** *Nitinol butterfly*

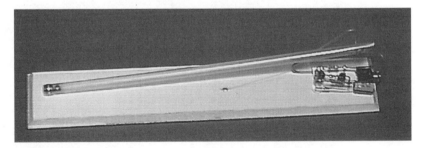

■ **4.4** *Rocker ball demo*

## Solenoids

Solenoids are electromechanical devices. A typical solenoid consists of a coil of wire that has a metal plunger through its center. When energized, the coil creates a magnetic field that either pulls or pushes the metal plunger (see Fig. 4.6). The metal plunger is mechanically connected to the robotic device that needs movement.

*Movement and drive systems*

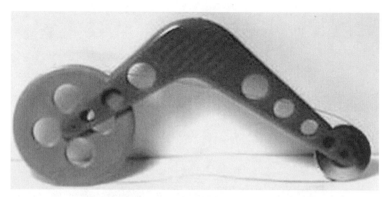

■ **4.5** *Heat engine*

■ **4.6** *Solenoid*

## Rotary solenoids

A rotary solenoid is a derivative of the standard solenoid (see Fig. 4.7). Instead of producing a linear motion, it produces a rotary motion. A rotary solenoid can be used to create a robotic fish (see Chap. 13).

■ **4.7** *Rotary solenoid*

■ **4.8A** *Stepper motor*

## Stepper motors

Stepper motors may be used for locomotion, movement, steering, and positioning control. These motors are used as integrated components in many commercial and industrial computer-controlled applications. For home personal computer (PC) users, stepper motors can be found in disk drives and printers.

Stepper motors are unique because they can be controlled using digital circuits. They are capable of precise incremental shaft rotation. This makes stepper motors ideal for rotary or linear positioning. Because stepper motors are widely used in industry, they come in a variety of shapes, sizes, and specifications (see Fig. 4.8A).

When power is applied to a standard electric motor, the rotor begins turning smoothly. The speed and position of the motor's rotor are a function of voltage, load on the motor, and time. Precise positioning of the rotor is not possible.

A stepper motor, however, runs on a sequence of electric pulses to the windings of the motor. Each pulse to a winding turns the rotor by a precise predetermined amount. The incremental movements of the rotor are often called *steps*. Hence the name, *stepper motors.*

Not all stepper motors rotate the shaft (rotor) by the same amount per step. They are manufactured with different degrees of rotation per step (or pulse). The optimum degrees per step will depend upon the particular application. Stepper motor specifications clearly state the degree of rotation per step. You can find a variety of stepper motors, with the range of rotation per step varying from a fraction of a degree (i.e., 0.72 degree) to many degrees (i.e., 22.5 degrees).

### Stepper motor circuit

Figure 4.8B is a schematic of a stepper motor driver circuit. The stepper motor in the circuit is a unipolar (six-wire) type. IC U1 is a 555 timer that is set up in astable mode to output square-wave clocking pulses on pin 3. U2 is a stepper motor controlling chip UCN 5804. The clocking pulses received on pin 11 of the UCN 5804 turn the stepper motor. Each pulse received on pin 11 turns the stepper motor one step. The faster the clocking pulses, the quicker the stepper motor turns.

In this sample circuit the clocking pulses are produced by a 555 timer. Clocking pulses can be generated by any number of sources like a microcontroller (discussed in Chap. 6) or a photoresistive neuron (discussed in Chap. 5). Switch SW1 is a fast/slow control. SW2 controls the stepper motor direction.

Stepper motors may be used in making a robotic platform (see Chap. 10).

## Servo motors

Servo motors are geared DC motors with positional control feedback. Hobbyist servo motors are commonly used for position control for radio-controlled (R/C) models. The shaft of the motor can be positioned or rotated through a minimum of 90 degrees.

Because of their widespread use in the hobby market, servo motors are available in a number of stock sizes (see Fig. 4.9). While larger

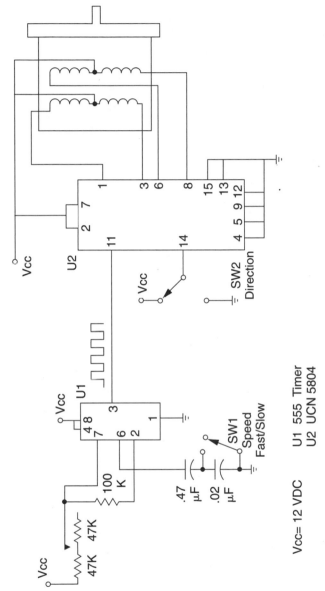

■ **4.8B** *Stepper motor*

Vcc= 12 VDC    U1  555 Timer
                U2  UCN 5804

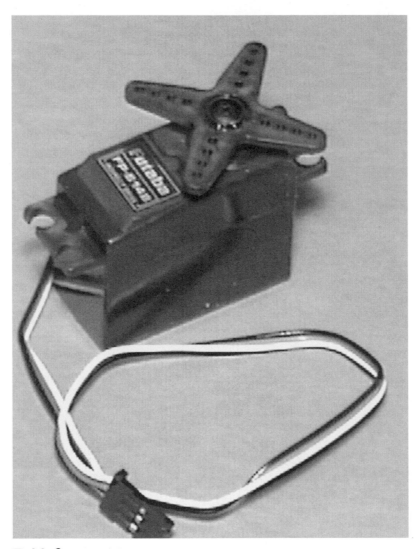

■ **4.9** *Servo motor*

industrial servo motors are also available, they are too expensive for most hobby applications. In this book we work with inexpensive and readily available hobby servo motors.

There are three wire leads to a servo motor. Two are for power, +4 to 6 V and ground. The third lead feeds a position-control signal to the motor. The control signal is a variable-width pulse between 1.0 and 2.0 milliseconds (ms). A neutral, midrange positional pulse is a 1.5-ms pulse. The pulse is sent 50 times a second (1 pulse every 20 ms or so) to the motor. This pulse signal will cause the shaft to locate itself at the midway position at ±45 degrees.

The shaft rotation on a servo motor is limited to approximately 90 degrees (±45 degrees from the center position). A 1-ms pulse will rotate the shaft all the way to the left (see Fig. 4.10), while a 2-ms pulse will turn the shaft all the way to the right. By varying the pulse width between 1 and 2 ms, the servo motor shaft can be rotated to any rotational degree position within its range.

You may feel that providing the pulse signal is a complex job; it isn't. The 16F84 PIC microcontroller, covered in Chap. 7, uses only a few lines of code to control a servo motor. And the PIC can control up to eight servo motors at a time. Another viable method is to utilize the servo motor control system used in R/C systems. Another alternative is to make your own circuit.

Making a servo motor circuit isn't as difficult as it may first appear. Figure 4.11 uses a 556 dual timer to control a servo motor. The 556 has two independent timers. To see the function more clearly, look at Fig. 4.12. Here two separate 555 timers are used. One timer is set in astable mode. The astable timer outputs a 55-hertz (Hz) square wave with a 1-ms negative component. The output from this timer is connected to the second 555 timer that is set up in monostable mode.

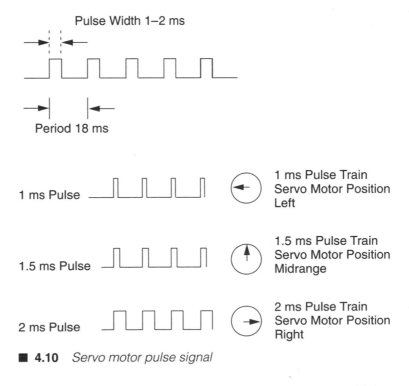

■ **4.10**   *Servo motor pulse signal*

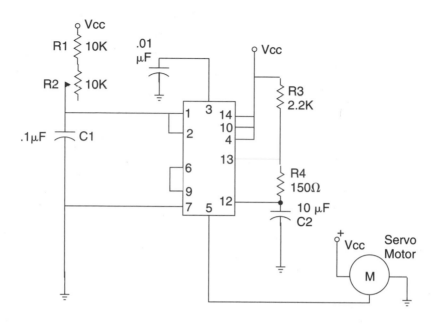

U1 = 556 Dual Timer

■ **4.11** *Servo motor circuit 556*

The monostable timer outputs a positive pulse from pin 5, each time it receives a negative pulse from timer 1. The positive pulse from timer 2 can be varied from 1 to 2 ms using the 10K-ohm potentiometer. You may need to fiddle with the resistance values of R1 and R2 in Fig. 4.11, depending upon the servo motor being used. As always, take caution that the servo motor isn't stalled (pushing against its own internal rotational stops).

When working with servo motors, I have found that to achieve full rotational movement from the shaft I needed to use pulses shorter than 1 ms and greater than 2 ms.

As you work with and gain experience using servo motors, you can also run outside the standard pulse widths (shorter and longer pulses) to achieve greater (180 degree) shaft rotation. The standard pulse widths sometimes do not rotate the shaft to each end stop.

However, before you do so you need to understand that if a pulse signal falls outside of the range where the servo motor shaft can rotate, the servo motor will fight against its own internal rotational stop in an effort to rotate the shaft to the position called for by the pulse.

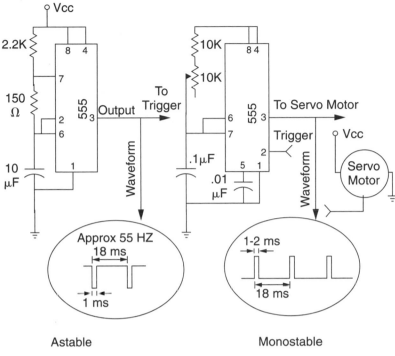

**■ 4.12** *Servo motor circuit 555*

For example, suppose you have a servo motor and you are providing a 2.8-ms pulse to rotate the shaft all the way to the right. As long as the shaft can rotate to that position, you're fine. But suppose a different servo motor reaches its maximum right rotation with a 2.5-ms pulse. If you start sending a 2.8-ms pulse width, you are ordering the servo motor to turn further than it physically can. Because the servo motor is fighting against the internal stop, current drain goes up, you increase wear on the gearing, and you may burn out the servo motor.

This problem usually occurs when the original servo motor tested is changed and the replacement motor doesn't need the modified pulse widths. As a rule of thumb, if using a pulse outside of the recommended 1- to 2-ms pulse, always check to make sure the servo motor isn't stalled in position.

Servo motors are used to make the walker robot in Chap. 11. The walker robot uses a PIC microcontroller to control the servo motors. The PIC microcontrollers and servo motor application are covered in Chap. 6.

## DC motors

DC hobby motors can be applied to movement and locomotion (see Fig. 4.13). Specifications of most DC motors show high revolutions per minute (rpm) and low torque. Robotics need low rpm and high torque. Gearboxes can be attached to motors to increase their torque while reducing the rpm (see Fig. 4.14). The gearbox usually specifies a ratio that describes the rpm in to the rpm out. For instance, a DC motor with an rpm of 8000 is connected to a 1000:1 gearbox. What is the output rpm? 8000 rpm/1000 = 8 rpm. The

■ **4.13**  *DC motor*

■ **4.14**  *DC motor with gearbox*

■ **4.15** *DC motor with gearbox head*

torque of the motor is substantially increased. You could estimate that the torque will increase by the same value the rpm decreased. In reality, no conversion is 100 percent efficient; there always will be efficiency losses.

Some DC motors, called *gearhead motors,* are built with a gearbox attached (see Fig. 4.15).

## DC motor H-bridge

When building a robot, one wants to control (turn on or off) the DC motor via a simple circuit or digital signal. In addition, one would also like to be able to reverse the motor's direction. An H-bridge fulfills these requirements.

It should be understood that the term "DC motor" refers to stand-alone DC motors as well as motors connected to gearbox motors as well as gearhead motors.

The H-bridge is made up of four transistors. [Some robotists use metal-oxide-semiconductor field-effect transistors (MOSFETs). I use NPN Darlington transistors.] Some H-bridge designers use a combination of PNP and NPN transistors. In each case, the transistor acts like a simple switch (see Fig. 4.16A). When switches SW1 and SW4 are closed (Fig. 4.16B), the motor rotates in one direction. When switches SW2 and SW3 are closed, the motor rotates in the opposite direction.

By using the switches properly, we can reverse the current direction to the motor, which in turn reverses the motor's shaft rotation. Figure 4.17 is an H-bridge circuit that uses transistors. An H-bridge circuit is used in Chap. 5 in the sensor tester robot.

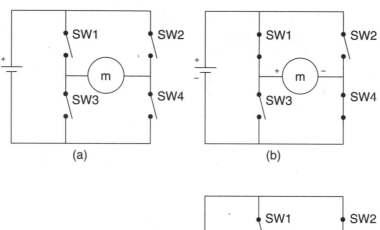

H-Bridge Function

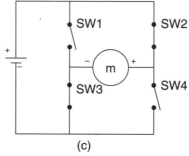

■ **4.16** *H-bridge using switches*

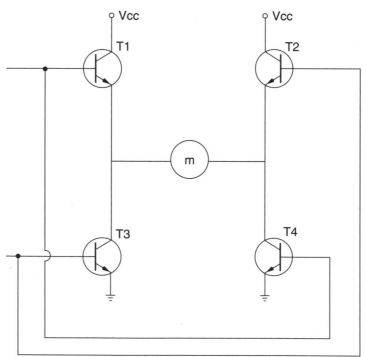

■ **4.17** *H-bridge using transistors*

## Pulse-width modulation

The H-bridge controls the on and off function as well as the direction of DC motors. The function of the H-bridge can be enhanced by using PWM to control the speed of the motor. The PWM signal is illustrated in Fig. 4.18. When the PWM signal is high, the motor is on; when low, the motor is off. Since the signal turns the motor on and off very quickly, the voltage delivered to the motor becomes an average of the time on versus the time period of the cycle (T-on/T-period). The greater the on time, the higher the average voltage. The average voltage (VDC steady-state) is always less than the voltage delivered (Vcc). PWM essentially controls the motor speed.

Motors are inductive loads. When current is switched on and off, a transient voltage is generated in the (motor) windings that can damage the solid-state components used in the H-bridge. This transient voltage can be controlled by using a snubber diode bridged across each transistor, as illustrated in Fig. 4.19.

The snubber diode dissipates the transient voltage by creating a voltage path directly to ground for the transient voltage. This effectively protects the semiconductor the diode is bridged over. The snubber diodes should be rated to handle the normal current the motor typically draws.

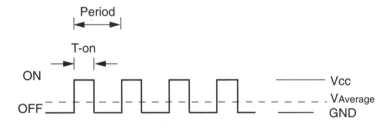

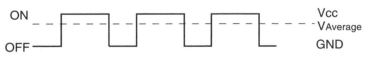

■ **4.18** *Pulse-width modulation for H-bridge*

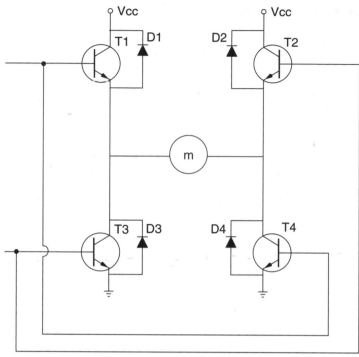

■ **4.19** *Transistor H-bridge with diode protection*

# Sensors

TYPICALLY ROBOTIC SENSORS MIMIC BIOLOGICAL SENSES like hearing, sight, touch, smell, and taste. Balance and body position derived from the inner ear are sometimes considered a sixth sense. Biological senses are neurally based, while robotic senses are electrically based. One could argue the point that they are both electrically based by pointing out that both the neural pathways and signals pass an electrochemical signal. However, neural sensors function differently than electrically based sensors. So, not to confuse technologies, it's important to define them as electrically based.

If one wants to truly imitate biological senses, neural sensors are needed. The human ear is an example of a neural sensor. Let's examine it. The human ear is not a linear instrument. Its response to sound is logarithmic. Because of this, a tenfold increase in sound intensity is only perceived by the human ear as a doubling of sound volume. In contrast, a common sound sensor, for instance, a microphone, has a linear response to sound intensity. Therefore, a tenfold increase in sound intensity is read by a computer (microcontroller or electronic circuit) as a tenfold increase in sound intensity.

Sensors detect and/or measure an aspect of the environment and can produce a proportional electrical signal. The signal information must then be read or interpreted by the intelligence [central processing unit (CPU)] or neural network on the robot. Although we may categorize the sensors as they relate to human senses, sensors are typically divided by the type of energy that the sensor responds to, such as light, sound, or heat. The sensors one incorporates into a robot will depend upon its intended operating environment and application.

## Signal conditioning

When determining which sensor to use for a robot, one must decide how the robot will read the sensor signal. Many sensors are resistance-type, meaning that the sensor varies its resistance in proportion to the energy being detected. When the sensor is placed in a simple voltage divider network, it outputs an electrical signal whose amplitude varies in proportion to the energy it senses.

If the robot is required to read the actual energy intensity (analog), an analog-to-digital (A/D) converter is needed. A/D converters can measure the electrical signal and output an equivalent binary number.

A/D converters will require a microcontroller or digital circuit to function properly and extrapolate the data. In many cases an A/D converter isn't required to read sensor signals. Instead of an A/D convertor one uses a comparator.

As its name implies, a *comparator* compares two voltages. One is a reference voltage that we, the designers, set. The other voltage is derived from the sensor (via the voltage divider). The comparator can output one of two signals: high or low. The high signal is +5 V and the low signal is 0 V.

The output signal from the comparator depends upon the magnitude of the two voltages on its two input lines. There are three possible choices. The sensor signal is less than the reference voltage, equal to the reference voltage, or greater than the reference voltage.

### Comparator example

The best way to learn about comparators is to use one in a circuit. The first thing you notice in Fig. 5.1 is that the comparator looks much like an operational amplifier (op-amp). This is true; comparators are specialized op-amps. The comparator used in our first example is the LM339 quad comparator. This integrated circuit contains four comparators in a 14-pin dip package. Like op-amps the comparator has an inverting and noninverting input. In this particular circuit the reference voltage is placed on the inverting input (−).

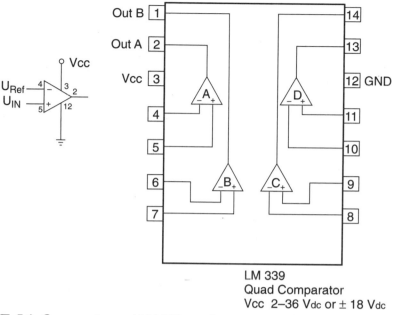

LM 339
Quad Comparator
Vcc  2–36 Vdc or ± 18 Vdc

■ **5.1** *Comparator and LM 339 quad comparator IC*

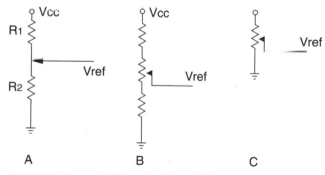

A           B           C

■ **5.2** *Voltage dividers A, B, and C*

## Voltage divider

The voltage divider is a simple but important concept. When using it, you will be able to connect most resistance-type sensors to a comparator. The reference voltage is derived from a voltage divider made up of two 10K-ohm resistors (see Fig. 5.2A). The Vref in this case will be half of the supply voltage (Vcc) of 5 V, or 2.5 V (see Table 5.1). We can make Vref any voltage we require between ground and Vcc by adjusting the two resistance values of the voltage divider.

$$\text{Vref} = \text{Vcc} \ \frac{R2}{R1 + R2}$$

where Vcc = 5 V.

To make an adjustable voltage divider, use a potentiometer as shown in Fig. 5.2B and C. I chose to use the Vref style of Fig. 5.2A because it is simple.

The schematic for our test circuit is shown in Fig. 5.3. In place of a sensor we will use two 1K-ohm resistors and one 5K-ohm potentiometer. By varying the potentiometer we can adjust the voltage going to the noninverting input (Vin). The output of each comparator is an uncommitted open collector of an NPN transistor. The transistor can sink more than enough current to light a

■ **Table 5.1 Two-Resistor Voltage Divider**

| R1 | R2 | Vref, V |
|---|---|---|
| 1K | 10K | 4.5 |
| 2.2K | " | 4.1 |
| 3.3K | " | 3.7 |
| 4.7K | " | 3.4 |
| 5.6K | " | 3.2 |
| 6.8K | " | 2.9 |
| 10K | " | 2.5 |

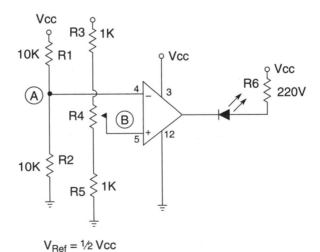

$V_{Ref} = \frac{1}{2} \ Vcc$

■ **5.3** Comparator test circuit schematic

light-emitting diode (LED), which we can use as an indicator. In addition, the output may be used as a simple single-pole, single-throw (SPST) switch to ground. This feature will be useful when we later need to trigger a 555 timer.

With the circuit wired, let's see what happens. When the input voltage (Vin) is less than the reference voltage (Vref), the output is 0 V (ground) and the LED is forward-biased and lit. If we adjust the potentiometer so that the voltage is greater than Vref, the output of the comparator goes high, turning off the LED. You can verify the operation of the comparator by using a voltmeter to measure the voltages at points A (Vref) and B (Vin).

Many people (myself included) feel this circuit is counterintuitive. I would like the LED to be lit when the sensor voltage is higher than the reference voltage. This can be accomplished by reversing the input leads and connecting the inverting input (−) Vin and the noninverting input to Vref. The output function reverses also.

When one doesn't need too many comparators, you may consider using a complementary metal oxide semiconductor (CMOS) op-amp configured as a comparator. The reason I like to use an op-amp is that it can source (supply) sufficient current to drive an LED or circuit directly (see Fig. 5.4).

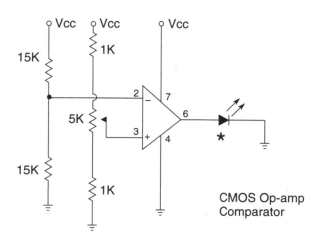

★ Sub miniature LED

■ **5.4** *Comparator op-amp test circuit schematic*

# Light sensors (sight)

There are a large variety of light sensors: photoresistive, photo-voltaic, photodiodes, and phototransistors. Light sensors can be used for navigation and tracking. Some robots use an infrared light source and detector to navigate around obstacles and avoid crashing into walls. The infrared source and detector are placed in front of the robot facing in the same direction. When the robot encounters an obstacle or wall, the infrared light is reflected off the surface causing an increase in the infrared light detected. The robot's CPU interprets this increased radiation as an obstacle and steers the robot around it.

Filters can be placed in front of light sensors to inhibit their response to some wavelengths while enhancing their response to others. One example of the use of filters is as flame detectors used in fire-fighting robots. One would try to enhance the response to light from a fire while inhibiting the response to light from other sources.

Another example is the use of colored gels as filters to promote color response. One could imagine a robot that separates or picks ripened fruit based on the fruit's skin color.

## Photoresistive

Cadmium sulfide (CdS) sensors (see Fig. 5.5) are photoresistors that can read ambient light. The CdS cells response to the light spectrum is in close approximation to that of the human eye (see Fig. 5.6). These are semiconductor sensors without the typical PN junction. The CdS cell displays its greatest resistance in complete darkness. As the light intensity increases, its resistance decreases. Measuring its resistance provides us with an approximation of ambient light.

## Photoresistive light switch

Figure 5.7 shows a basic light switch. Because the CdS cell is a resistive-type transducer, it can be placed as a resistor in a voltage divider. When the light intensity increases, the resistance of the

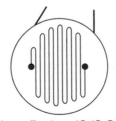

Photo-Resistor(CdS Cell)     ■ **5.5** *Cadmium sulfide cell*

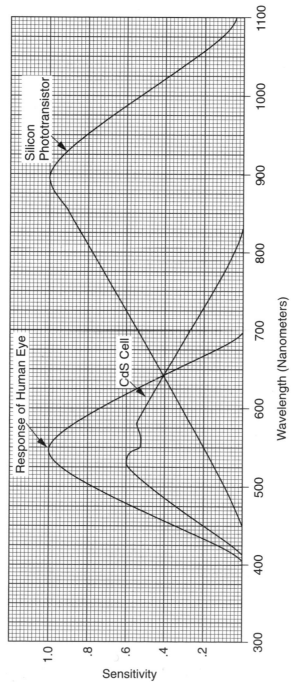

■ **5.6** *Spectrum chart showing response of eye, CdS cell, etc.*

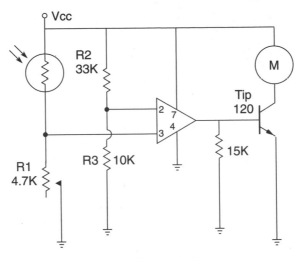

■ **5.7** *Photoresistive light switch*

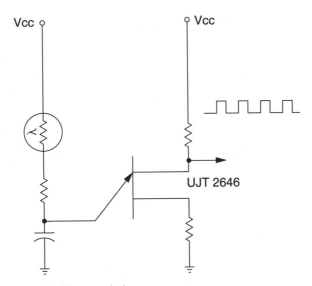

■ **5.8** *Photoresistive neuron*

CdS cell decreases. This increases the voltage drop on R1 and is seen on pin 2. When the voltage on pin 2 is greater than the voltage on pin 3, the motor turns on. The threshold is adjusted using R1, a 4.7K-ohm personal computer (PC)–mounted potentiometer. This is the basic circuit that controls the solar ball project in Chap. 12.

## Photoresistive neuron

Figure 5.8 shows a light neuron. As the intensity of light increases, the pulse rate becomes faster. The light neuron can provide the clock pulses to a stepper motor controller chip like the UCN5804. As the light increases, the stepper motor turns faster.

## Photovoltaic

Solar cells, photodiodes, and phototransistors are similar in construction. They all have a light-sensitive PN junction. Solar cells use a wide-area PN junction to produce electric power in proportion to light intensity.

Photodiodes are usually reverse-biased in a circuit. When light strikes the diode's PN junction, the current flows. The photodiode has a quicker response than the CdS cells and can relay information encoded in the light.

Phototransistors are light-sensitive transistors. The advantage of a phototransistor over the photodiode is that it can provide amplification of the light signal.

## Infrared

Infrared (IR) sensors detect low-frequency [900-nanometer (nm) and longer] light. They deserve special consideration because they are widely used in robotics for tracking, collision avoidance, and communication.

Using infrared sensors has never been easier. Infrared receiver modules that incorporate modulation detection, shown in Fig. 5.9, are available through a number of electronic distributors. The

■ **5.9** *Infrared receiver module*

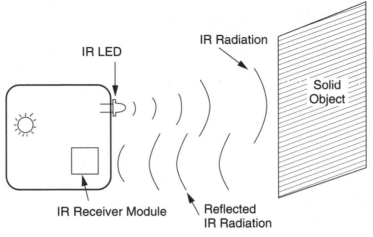

**IR LED**

**IR Radiation**

**Solid Object**

**IR Receiver Module**

**Reflected IR Radiation**

■ **5.10** *Drawing of infrared collision detector*

advantage to these modules is that they only detect IR light oscillating at a specific frequency [usually around 40 kilohertz (kHz)].

The 40-kHz waveform can be modulated by another (lower-frequency) signal. The receiver module has also been designed to receive an impressed signal on the 40-kHz carrier wave. This produces a robust communication link. Primarily the receiver module responds only to the 40-kHz IR signal, permitting the receiver to "see" the IR light being transmitted from the transmitter, reject other light sources, and detect the modulation on the 40-kHz wave.

### Infrared collision detector

Figure 5.10 is a drawing of a simple collision detector. As the sensor approaches a solid object, the IR light reflected back into the receiver increases. The increased IR light reaches a specific amplitude where it trips a comparator circuit informing the robot there's an obstacle ahead.

### Infrared transmitter

Figure 5.11 is a schematic of the IR transmitter. The transmitter uses a 555 timer set up in astable mode. Potentiometer R1 is used to adjust the frequency output. The output of the timer (pin 3) is connected to a 2N2222 NPN transistor. An infrared LED is connected to the emitter of the transistor. When you turn on the circuit, don't expect to see any light being emitted from the LED. The infrared light is not detectable by the human eye. Because we are using

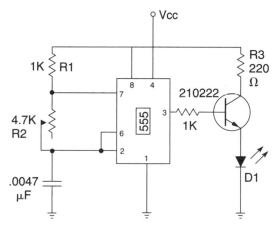

■ **5.11** *Schematic of infrared transmitter*

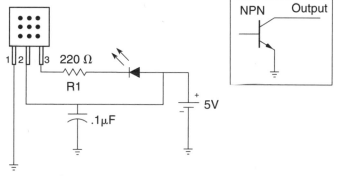

■ **5.12** *Schematic of infrared receiver*

this in a simple collision detector, there is no need to modulate the 40-kHz signal.

### Infrared receiver

Figure 5.12 is a schematic of the infrared receiver. The receiver module is an Everlite IRM-8420. The center frequency is 37.9 kHz with a bandwidth of 3 kHz (±1.5 kHz). The output is active low. This means that when the receiver module detects the signal, the output drops to ground. The output is equivalent to an open collector of an NPN transistor (see the insert of Fig. 5.12). The output can sink sufficient current to light an LED. In the test circuit the LED will light when the module is receiving the signal.

**Tuning the transmitter**　Set up the infrared diode and receiver module next to one another facing in the same direction. The LED must be completely encased in a tube of some sort that only permits the infrared light to leave from the front of the LED. Failure to do this will make using this setup impossible. Note that some plastic materials while opaque to visible light are completely transparent to infrared light.

Place a white card about 3″ in front of the transmitter and receiver. Turn on the circuit. Adjust R1 until the receiver's LED turns on. Then remove the white card. The receiver's LED should go off. If it doesn't, the infrared LED on the transmitter may be leaking light from the side and activating the receiver.

Once the unit is working properly, the circuit can be fine-tuned to detect objects at a greater distance. Move the white card back in front of the transmitter and receiver until it just triggers the LED to turn on. Adjust the potentiometer (slightly) on the transmitter so that the LED turns on completely. Keep in mind that it may not be advantageous for the robot to detect objects and/or collisions that are too far away.

## DTMF IR communication/remote control system

Other authors have detailed the use of IR transmitters for communication and remote control. Typically the IR transmitter is modulated at a particular frequency and the receiver unit utilizes a 567 phase-locked loop (PLL) integrated circuit (IC). While this works, one must match and tune each transmitter-receiver pair. There is an acceptable way to work around this.

Integrated circuit chips designed and manufactured for the telecommunications industry are readily available. These inexpensive chips are capable of transmitting and receiving 16 distinct signals, no tuning required. By coupling these chips to standard IR components, an IR remote communication/control system can be implemented.

## DTMF

The dual-tone multifrequency (DTMF) signal was originally developed just over 25 years ago. This was before the U.S. government forced Bell Telephone to break up, allowing the company to expand into other markets. DTMF is commonly known as touch-tone dialing.

The standard DTMF signal is composed of two audio tones generated from a group of eight possible tone frequencies. The eight frequencies are divided into two equal groups, a low-frequency group and a high-frequency group (see Table 5.2). The DTMF signal is an algebraic sum of two tone frequencies, one tone from each frequency group (see Figs. 5.13 through 5.15). If we do the math, we see that there are $4 \times 4 = 16$ possible combinations.

■ Table 5.2

| Pin# | Row# or column# | Frequency, Hz |
|------|-----------------|---------------|
| **Low-frequency group** | | |
| R1 | Row 0 | 697 |
| R2 | Row 1 | 770 |
| R3 | Row 2 | 852 |
| R4 | Row 3 | 941 |
| **High-frequency group** | | |
| C1 | Column 0 | 1209 |
| C2 | Column 1 | 1336 |
| C3 | Column 2 | 1477 |
| C4 | Column 3 | 1633 |

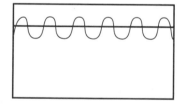

■ 5.13 *Low-frequency tone waveform*

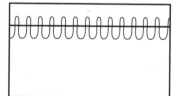

■ 5.14 *High-frequency tone waveform*

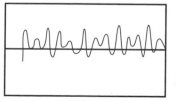

■ 5.15 *Algebraic sum of low and high frequencies (DTMF)*

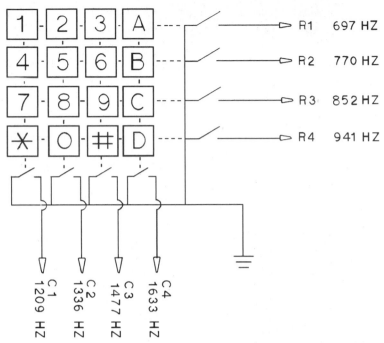

■ **5.16** *4 × 4 keypad matrix showing individual DTMF frequencies*

The low frequencies (R1 to R4) are referred to as the *row group*. The high frequencies (C1 to C4) are referred to as the *column group*.

### DTMF encoding

Any combination of frequencies can be obtained using a 4 × 4 matrix of switches or keypad (see Fig. 5.16). Remember we are borrowing this technology from the telephone industry; it has been designed for optimum efficiency for less-than-perfect telephone lines.

Standard touch-tone telephones use a 3 × 4 keypad matrix. This switch matrix provides coding for all the row frequencies and only three column frequencies (see Fig. 5.17). A 3 × 4 keypad matrix is more readily available and has been used with all the circuits described here.

Not all telephone keypads are made the same; therefore, some keypads on the market will not be suitable in these circuits. For instance, some keypads have different internal switch wiring or include proprietary ICs. So if you try another keypad, keep that in mind if the circuit fails to operate properly.

Building a DTMF encoder is simple (see Fig. 5.18). The circuit only requires a keypad, crystal, and 5089 IC. The pin out of the

# KEYPAD WIRING

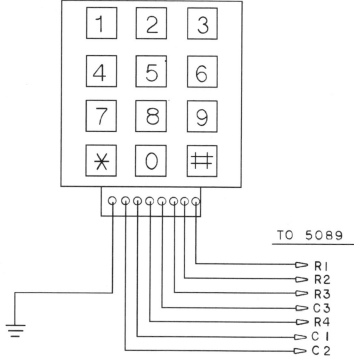

TO 5089

R1
R2
R3
C3
R4
C1
C2

■ **5.17** *Wiring 3 × 4 telephone keypad*

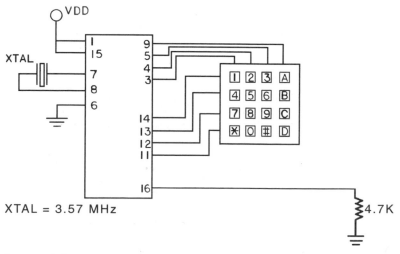

VDD

XTAL

XTAL = 3.57 MHz

4.7K

■ **5.18** *DTMF encoder circuit using 4 × 4 keypad matrix*

5089 is shown in Fig. 5.19. If you use a standard 3 × 4 (telephone) keypad, you will lose the four functional DTMF codes associated with the missing keys, therefore reducing the maximum number of usable channels to 12.

Figure 5.20 is an encoder test circuit that uses an eight-position dip switch. The dip switch takes the place of the matrix keypad; with it you can test the operation of this encoder circuit and the receiver (decoder) circuit. Notice when you turn a switch on, you are grounding the pin it is connected to. Pins R1 through R4 and C1 through C4 are active low. Dip switches 1 through 4 are connected to pins R1 through R4, and dip switches 5 through 8 to pins C1 through C4.

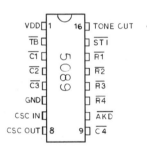

Pin out of 5089 DTMF Transmitter

| PIN# | Name | Function |
|------|------|----------|
| 1 | VDD | Supply Voltage +5V |
| 2 | *TB | Tone disable |
| 3 | *C1 | Column input |
| 4 | *C2 | Column input |
| 5 | *C3 | Column input |
| 6 | Vss | Ground |
| 7 | OSC1 | Clock input |
| 8 | OSC2 | Clock input |
| 9 | *C4 | Column input |
| 10 | *AKD | Any Key Down output |
| 11 | *R4 | Row input |
| 12 | *R3 | Row input |
| 13 | *R2 | Row input |
| 14 | *R1 | Row input |
| 15 | STI | Single tone inhibit input |
| 16 | Tone O/P | DTMF output |

\* Active Low

■ **5.19** *Pin out of 5089 DTMF encoder IC*

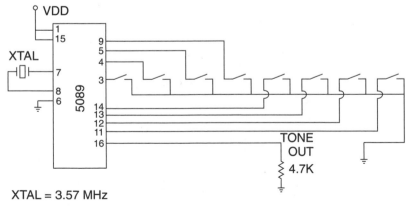

XTAL = 3.57 MHz

■ **5.20** *Schematic of encoder circuit using eight-position dip switch*

Pin Out of G8870 DTMF Receiver

| PIN# | Name | Function |
|------|------|----------|
| 1 | IN+ | Noninverting Input |
| 2 | IN- | Inverting Input |
| 3 | GS | Gain Select |
| 4 | Vref | Reference Voltage |
| 5 | IC | Internal connection |
| 6 | IC | Internal connection |
| 7 | OSC1 | Clock input |
| 8 | OSC2 | Clock input |
| 9 | Vss | Ground |
| 10 | TOE | Tri-state enable output |
| 11 | Q1 | Output |
| 12 | Q2 | Output |
| 13 | Q3 | Output |
| 14 | Q4 | Output |
| 15 | StD | Delay steering output |
| 16 | ESt | Early steering output |
| 17 | St/GT | Steering input/guard time |
| 18 | VDD | Supply Voltage.+5V |

■ **5.21** *Pin out of G8870 DTMF decoder IC*

The IC can also produce single tones. These are usually generated for testing purposes. For instance, to generate a 1336-hertz (Hz) tone equivalent to that of the C2 pin, ground any two row pins and the C2 pin. This action will generate a single 1336-Hz signal. The same may be done with the row frequencies. Ground any two column pins with the particular row frequency pin you want to generate.

### DTMF decoding

DTMF decoding is just a little more complex than encoding. Again the simplicity results from the use of a single IC chip, in this case the G8870 (see Fig. 5.21).

The decoding chip has a 4-bit latched output labeled Q1 through Q4. Q4 is the most significant bit (MSB). The current available from the outputs of Q1 through Q4 is sufficient to light a low-current LED. Figure 5.22 is a basic receiving circuit. The output from Q1 through Q4 lights the LED and is a binary number. By looking at Table 5.3, you can determine the binary output that will be displayed on the Q1 through Q4 for all DTMF signals. The way the circuit is wired, the binary "1" will be represented by a lit LED.

### Microcontroller

The 4-bit number from the G8870 can be connected directly to input lines of a microcontroller like the PIC 16F84. The microcontroller can easily read this binary number. We will get to the PIC microcontrollers in Chap. 7.

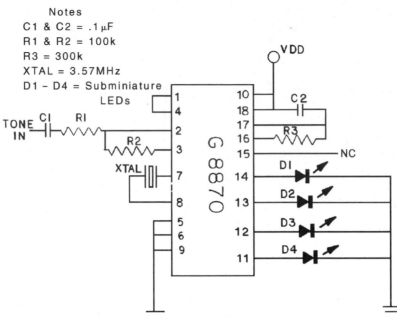

Notes
C1 & C2 = .1 μF
R1 & R2 = 100k
R3 = 300k
XTAL = 3.57MHz
D1 – D4 = Subminiature LEDs

■ **5.22** *Schematic of receiver circuit with 4-bit binary output*

■ Table 5.3  DTMF Output

| | | Signal frequency, Hz | |
| --- | --- | --- | --- |
| Decimal | Binary | Low | High |
| 1 | 0001 | 697 | 1209 |
| 2 | 0010 | 697 | 1336 |
| 3 | 0011 | 697 | 1477 |
| 4 | 0100 | 770 | 1209 |
| 5 | 0101 | 770 | 1336 |
| 6 | 0110 | 770 | 1477 |
| 7 | 0111 | 852 | 1209 |
| 8 | 1000 | 852 | 1336 |
| 9 | 1001 | 852 | 1477 |
| 0 | 1010 | 941 | 1336 |
| * | 1011 | 941 | 1209 |
| # | 1100 | 941 | 1477 |
| A | 1101 | 697 | 1633 |
| B | 1110 | 770 | 1633 |
| C | 1111 | 852 | 1633 |
| D | 0000 | 941 | 1633 |

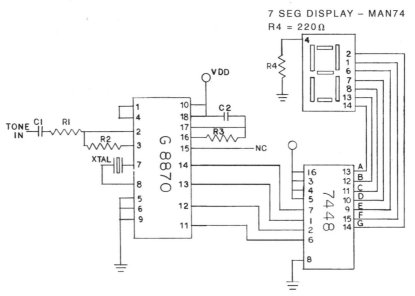

**■ 5.23** *Schematic of receiver circuit with digital display*

The IR link discussed in the next few paragraphs combined with the PIC microcontroller outlined in Chap. 7 will allow users to program communications between mobile robots for games like tag and follow the leader.

### Adding a digital display

If reading binary numbers is too cumbersome, we can add a digital numerical display. The output from the chip may also be fed to a binary-coded-decimal (BCD) to 7-segment decoder chip, such as the 7448. The 7448 IC is connected to a 7-segment display like the MAN 74 (common cathode). These two chips will provide a digital readout (see Fig. 5.23).

**Testing**   For testing purposes connect the output from the 5089 chip (pin 16) to the input of the G8870 chip, using either a keypad or dip switches to generate the DTMF signals. The receiver will display the output via the LEDs or segmented display.

### Adding IR transmission

Once the DTMF chips are operating properly, it becomes a simple matter to connect the chips via an IR link. The output of the 5089 chip is connected to the base of a common NPN transistor (see Fig. 5.24). A high-power IR LED diode is connected to the emitter

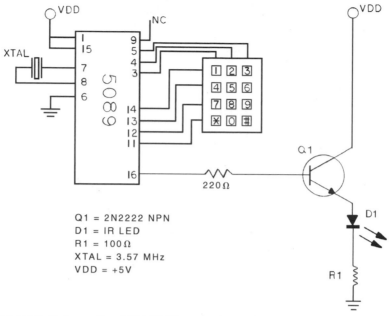

**■ 5.24** *Schematic of IR DTMF transmitter circuit*

of the transistor. Although the IR LED may be connected directly to the output of the 5089, the power output would be small. The NPN transistor allows additional current to power the LED.

Figure 5.25 shows the front end of the IR receiver. An IR phototransistor is coupled to a CMOS op-amp. This combination of components allows the receiver chip (8870) to lock in on the IR radiation from a distance of several feet.

### Remote control

Using the IR link, you should be able to press a number on the keypad and see the corresponding number displayed on the digital display. Test the IR link at this point for maximum distance and direction. You should be able to increase the distance by placing the IR LED and phototransistor in their own reflectors. The light reflector from an old flashlight will work well.

The remote control begins by adding a 4028 IC. The 4028 is a BCD-to-decimal decoder, meaning it reads the binary number (remember the four LEDs from Fig. 5.22) and outputs a single line equal to the decimal equivalent. The 4028 has 10 (0 to 9) output lines. Whatever 4-bit binary number is placed on its input lines, the 4028 outputs a high signal on that output line (see Fig. 5.26).

It is not necessary to remove the 7448 and 7-segment display. The 8870 chip has sufficient output to drive both the 7448 and 4028. The digital display is pretty handy when checking the output from the 4028. For the sake of simplicity, Fig. 5.26 just shows the 4028 connected to the 8870.

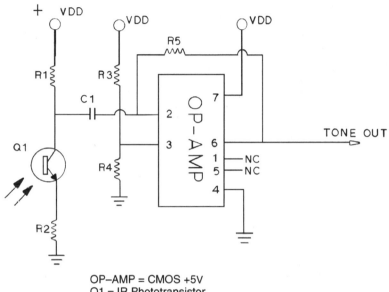

OP–AMP = CMOS +5V
Q1 = IR Phototransistor
R1 = 1K
R2 = 100Ω
R3 & R4 = 10K
R5 = 100K

■ **5.25** *Schematic of front end of IR DTMF receiver*

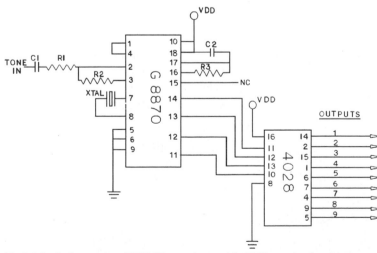

■ **5.26** *Schematic of DTMF receiver with BCD-to-decimal converter*

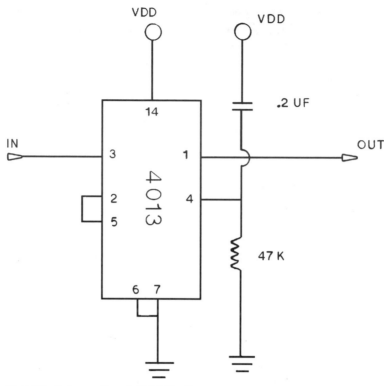

■ **5.27** *Schematic of 4013 flip-flop*

The output from the 4028 can be used directly to turn a switch or circuit on or off. However, this isn't an optimum situation, because as soon as you key another number (channel), the previous channel turns off (brings the line low).

The solution to this problem is a 4013 D-type flip-flop (see Fig. 5.27). The flip-flop is a basic computer memory datum. In this circuit it is configured as a divide-by-two counter. Upon receiving the first "on" signal from the 4028, it turns its output line high. When the 4028 brings the line low, which happens when hitting another channel, the 4013 will keep its output line high (latched).

To bring the 4013 output line low, simply key the channel for a second time. The second high signal to the 4013 brings the output line low (unlatched). One can continue to bring the 4013 output line high and low by alternately switching the input line high.

## Machine vision

To reproduce human vision in a machine is a difficult task. One cannot simply connect a video camera to a computer and expect it to see. Programs (both neural and expert) must capture the video image and process it (extrapolate data). Machine vision has been achieved in limited and targeted areas.

Chapter 1 looked at the Papnet computer, which uses neural software to analyze pap smear slides with a higher accuracy than can be achieved by humans. Other researchers have developed vision systems that can steer a vehicle based on the contours of the road being driven.

Before we can attempt to simulate human vision, we need (in addition to developing improved image processing, which is no easy task in itself) to develop stereoscopic mounted video cameras. Some research in this area is taking place at the Massachusetts Institute of Technology (MIT) on their humanoid robot, COG. With stereoscopic cameras, two video pictures must be processed and then merged to create a three-dimensional (3D) representation. This is the same process used in human 3D vision. To estimate depth, each camera must be mounted on gimbals that allow the cameras to veer in (converge) and focus on an object. The amount of convergence is taken into consideration for judging the distance of objects.

Machine vision is a fertile field of development. Currently most vision systems require a high-powered computer dedicated just to vision processing.

## Body sense

Body sense provides some information on where one is and what position one is in. Limited body sense can be accomplished in robots by using a variety of tilt switches (see Fig. 5.28). This will at least

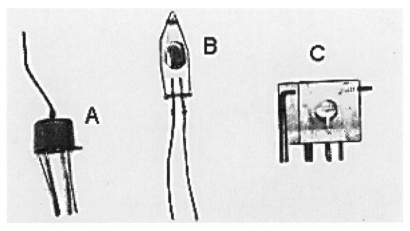

■ **5.28** *Tilt switches*

inform a robot if it is on an incline or decline, flat on its back or on its stomach, or upside down or right side up. The robot can then take appropriate action based on its body sense to accurately change position.

## Direction—magnetic fields

Using the Earth's magnetic field, an electronic compass can provide directional information. This will allow a robot to travel in a certain direction or to know which direction it's traveling in.

The simplest sensor in this category is the 1490 digital compass (see Fig. 5.29). The compass is a solid-state Hall device. The digital compass provides four outputs that represent the four cardinal directions: north, east, south, and west. Using a little logic a total of eight directions can be determined.

The compass is dampened to approximate the speed of a liquid-filled compass. It takes 2.5 seconds (s) for it to respond to a 90-degree displacement. The damping prevents overswinging the direction and prevents switch fluttering when near a switching direction. The device is sensitive to tilting. Any tilt greater than 12 degrees will cause directional errors.

The bottom of the device has 12 leads arranged in four groups of three. Looking at the device from the top, the leads in each group are labeled 1, 2, and 3. The leads labeled 1 are connected to Vcc (+5V). The leads labeled 2 are connected to ground. The leads labeled 3 are the four outputs. The outputs of the digital compass are equivalent to open collectors of an NPN transistor. Being open collects, the outputs are unable to source any current but are capable of sinking enough current [20 milliamps (mA)] to light LEDs.

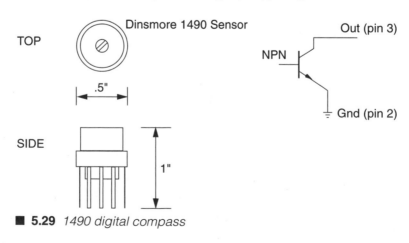

■ **5.29** *1490 digital compass*

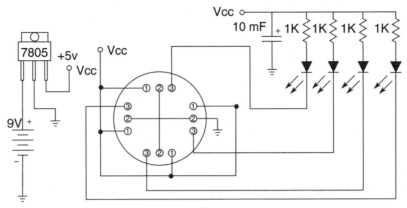

■ **5.30** *Digital compass test circuit using four LEDs*

The test circuit is shown in Fig. 5.30. The sensor will operate with supply voltages ranging from +5 to 18VDC. A 9V battery is used as a power source and is regulated to +5V using a 7805 voltage regulator.

As a rule, try to keep all voltages at 5 V maximum. This will make it computer-safe. For instance, when the digital compass is interfaced to the PIC microcontrollers, if we forgot and used a 9V power source on the compass, the outputs may damage the input/output (I/O) inputs.

The test circuit uses four LEDs for display. As the sensor is rotated, each cardinal position on the compass will light one LED. The intermediate directions light two LEDs.

### Testing and calibration

Find north using a standard compass. Rotate the circuit so that one LED is lit. I used the LED furthest away from the sensor for north. If you do the same, the other LEDs will automatically follow the same sequence outlined (see Fig. 5.31). The sequence for my display is as follows: 1 = on, 0 = off.

### Computer interface

The four output lines from the compass form a 4-bit binary number (nibble) that is easily read by a microcontroller, computer, or electronic circuit (see Table 5.4). We will hold off on the PIC microcontroller circuit until we have introduced the 16F84 PIC microcontroller in Chap. 7.

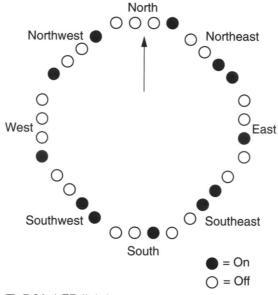

North

Northwest ● ○ Northeast

West ○ ○ East

Southwest ● ○ Southeast

South

● = On
○ = Off

■ **5.31** *LED lighting sequence*

■ Table 5.4 **LED Sequences**

| Direction | LEDs | Decimal equivalent | Inverted |
|-----------|------|--------------------|----------|
| North | 0001 | 1 | 14 |
| Northeast | 0011 | 3 | 12 |
| East | 0010 | 2 | 13 |
| Southeast | 0110 | 6 | 9 |
| South | 0100 | 4 | 11 |
| Southwest | 1100 | 12 | 3 |
| West | 1000 | 8 | 7 |
| Northwest | 1001 | 9 | 6 |

## 1525 electronic analog compass

In most cases the 1490 directional information is more than suffi-cient for a robot. However, there will be cases when high-resolu-tion directional information may be important. In this case one may use the 1525 electronic compass (see Fig. 5.32).

The signal output from the 1525 is much harder to read than that of the 1490, but the tradeoff is that the 1525 electronic compass provides a directional resolution of approximately 1 degree.

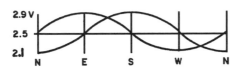

TYPICAL SIN-COS COMPASS RESOLUTION

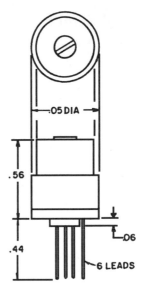

■ **5.32** *1525 electronic analog compass*

The output of this compass is comprised of two sine waves, one of which is 90 degrees out of phase with the other (see the sine-cosine configuration in Fig. 5.32). The amplitude of each wave correlates to direction. If a 90-degree portion of a sine wave is measured with an 8-bit A/D converter, a compass directional resolution of 1 degree is obtained.

## GPS

Using a global positioning system (GPS), a robot can know precisely where on Earth it's located. While the need for a GPS is not obvious for amateur robotists, the cost of GPS systems are coming down if the need arises.

## Speech recognition

The human ear has an auditory range from 10 to 15,000 Hz. Sound can be picked up easily using a microphone and amplifier. Microphones typically have an auditory range that surpasses that of human hearing. Sound is a useful tool for robotists.

We use hearing primarily for communication (language). Speech-recognition systems are a hot topic in robotics. Because of this, we devote an entire chapter (see Chap. 7) to building a speech-recognition circuit and interfacing. But don't skip over the following information. Robotic sound systems are pretty useful.

## Sound and ultrasonics

Sound may be used for games, range finding, and collision and obstacle avoidance. To play a game of robot tag, robots are fitted with a two-tone oscillator and receiver. Each robot can generate and recognize two tones. Let's say the A tone is 3000 Hz and the B tone is 6000 Hz. Tones are generated for 1 s whenever a robot's bumper switch is activated.

The robot that's "it" generates the B tone whenever its bumper collides with another robot. A "not it" robot generates the A tone. Upon collision, the "it" robot generates the "B" tone. The "not it" robot hearing the B tone changes states and becomes "it." And the "it" robot hearing the A tone from the "not it" robot changes states and becomes "not it." Two "not it" robots will both generate the A tone and leave the collision with their states unchanged. Although we are using sound as an example here, be aware that this technique can be applied using infrared light.

Ultrasonics are often used for range finding and collision detection. Many robotists have written on the Polaroid Company's ultrasonic modules (see Fig. 5.33). These modules are used in Polaroid cameras to quickly measure the subject's distance from the camera and focus the lens to produce sharp pictures. When interfaced to a microcontroller, the units can accurately measure distance.

If one needs or wants distance measurements for the robot, the Polaroid sensor is the way to go. The ping can measure distances up to 30 ft. The sensor may also be rotated (using a servo or stepper motor) like a radar to build a navigation map and find an obstacle-free path.

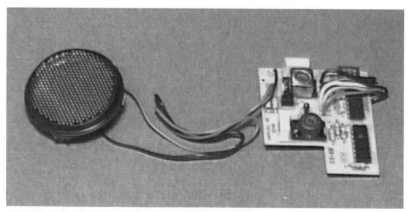

■ **5.33** *Polaroid ultrasonic ranging module*

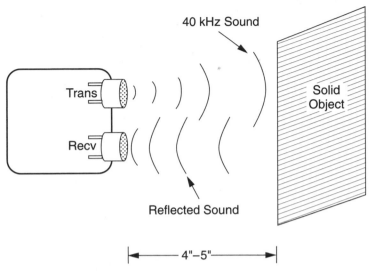

40 kHz Sound

Trans

Recv

Solid
Object

Reflected Sound

|← 4"–5" →|

■ **5.34** *Outline of ultrasonic collision detector*

Every time the Polaroid transducer is energized, there is an audible click from the Transducer. I find the constant clicking from the sensor annoying. Although the module is ultrasonic, when the electronics pump the ultrasonic signal to the transducer, some audible sound is also generated.

It is relatively easy to build a basic ultrasonic collision avoidance system that, being completely ultrasonic, is silent. The basic operation follows the same scheme used for infrared collision avoidance, except we are using sound instead of light. Figure 5.34 shows the overview schematic. The transmitter sends a 40-kHz signal to an ultrasonic transducer. Another transducer (receiver) is positioned alongside the transmitter transducer. When the robot approaches a wall or obstacle, the 40-kHz sound is reflected back to the receiver, whose output increases in amplitude. When the output increases beyond the preset point, the comparator trips, relaying that there is an obstacle detected.

## Ultrasonic receiver section

The ultrasonic receiver section (see Fig. 5.35) is used to fine-tune the transmitter. The ultrasonic transducers resonate at 40 kHz. If the resonant frequency varies too much (±750 Hz), the performance of the transducers degrades rapidly. Fine-tuning the transmitter for optimum resonance frequency is not difficult provided you follow the procedure outlined. The only piece of equipment needed is a volt-ohm milliammeter (VOM) capable of reading 2 VDC.

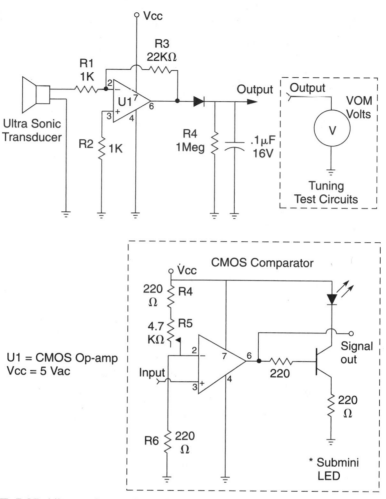

■ **5.35** *Ultrasonic receiver circuit*

Because the transducers have a limited bandwidth (resonant at or around 40 kHz), it is unnecessary to add a PLL (LM567) to the circuit. The transducers naturally reject off-frequency sound.

The receiver section uses a CMOS op-amp. The op-amp is an 8-pin dip that follows the same pin out as the universal 741 op-amp. (Do not substitute a 741 op-amp.) The op-amp is configured as an inverting amplifier with a gain of approximately 22.

## Ultrasonic transmitter section

The ultrasonic transmitter (see Fig. 5.36) is built around a CMOS 555 timer set up in astable mode. R2 is a PC-mounted 4.7K-ohm potentiometer and is used to adjust the frequency output.

### Tuning the transmitter

Set up the ultrasonic transducers so that they are directly facing one another about 4" to 5" apart (see Fig. 5.37). Connect the VOM to the circuit as shown in the insert in Fig. 5.35. (Leave the comparator section off.) Set the VOM to read volts DC. You will need to read about 2 V; set the range on the VOM accordingly. Turn on both circuits. Adjust R2 of the transmitter so that you obtain the peak voltage output shown on the voltmeter. This should read about 2 VDC.

### Adjusting the CMOS comparator

After tuning the transmitter, we need to set the receiver's comparator circuit. Disconnect the VOM from the receiver section and connect the CMOS comparator. Rearrange the transducers so that they are lying side by side about half an inch apart facing in the same direction. Place a flat-sided solid object about 3" in front

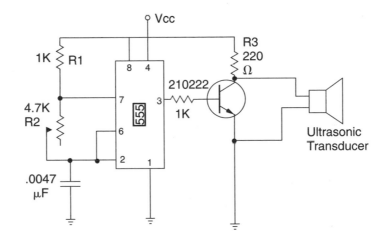

Vcc = 5Volts

\* Use CMOS 555 Timer

■ **5.36** *Ultrasonic transmitter circuit*

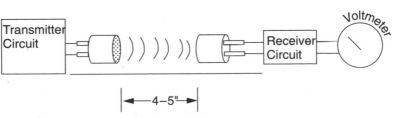

■ **5.37** *Ultrasonic test setup*

of the transducers. Turn on the receiver and transmitter circuits and adjust R5 on the receiver circuit so that the subminiature LED just lights.

To test the circuit, remove the solid object from the front of the transducers; the LED should turn off. Fine-tune the circuit by placing the solid object 5″ to 6″ in front of the transducers and readjust R5 until the LED just lights. Note that the receiver is angle sensitive. If the object is held at an acute angle, the ultrasonic sound is reflected away from the receiver. The angles become less critical as the object gets closer to the transducers.

The circuit easily detects solid objects up to 8″ away from the transducers. Greater distances can be detected, but as mentioned earlier they become angle sensitive. I have the transducers set perpendicularly. You may angle the transducers slightly to obtain different ranging effects.

The circuit provides a transistor-transistor logic (TTL) high signal that is indicated by the lit LED whenever the circuit detects an obstacle 6″ away. The TTL signal may be read directly by a neural net or microcontroller.

## Arranging the ultrasonic sensors

The obvious uses for the ultrasonic system are side (left and right), front, and back obstacle detection. Another use that may not be as obvious is ground detection. If the ultrasonic sensor faces forward and is pointed downward, the sensor will read the ground in front of the robot. If the robot approaches a cliff or stair, the normally high signal (LED lit) goes low informing the CPU to stop.

# Touch and pressure

The fidelity of the human sense of touch has not been remotely approached in robotics. However, there are a few simple sensors that can be used to detect touch and pressure. Touch sensors are commonly to detect bumps in the robot's path and to allow the robot to avoid collisions.

More sophisticated touch and pressure sensors are used on robotic hands and arms. The sensors allow the robotic hand to grip with enough force to lift an object without crushing it.

A simple touch or pressure sensor can be made from electrostatic (also called conductive) foam. This is the same foam ICs are

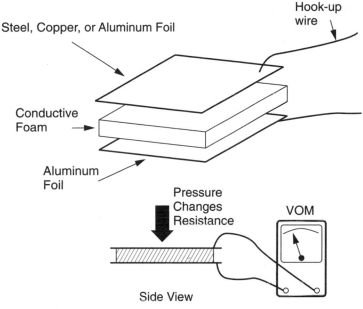

**■ 5.38** *Conductive foam touch sensor*

packed in to prevent static damage. The foam has a nominal conductivity that changes as the material is compressed.

It is important to use low-density (soft) conductive foam, because it is soft and spongy. As pressure is applied, the foam compresses, which changes the nominal resistance between the conductors.

Figure 5.38 illustrates a simple touch sensor. The conductive plates may be made from printed circuit flexible board (PCB), aluminum foil, or something similar. Higher-fidelity touch and pressure sensors are reviewed a little later in this chapter.

## Piezoelectric material

There are a great many piezoelectric sensors. Piezoelectric sensors can detect vibration, impact, and thermal radiation. Pennwall Company makes a unique product called piezoelectric film. This is an aluminized plastic that's been manufactured in such a way as to render the plastic piezoelectric.

The material is sensitive enough to detect the thermal radiation of a person passing in front of it. Many commercial light sentries sold in hardware stores use piezoelectric film behind a Fresnel lens to detect the thermal radiation of a person. This type of light sentry automatically turns on a light when someone walks into its field of view.

## Switches

Momentary contact switches form the foundation of bump sensors, navigation feelers, and limit sensors. There are many types of switch configurations to choose from. Some of the more common switches used in robotics are momentary contact lever and pushbutton switches (see Fig. 5.39).

## Bend sensors

Bend sensors are passive resistive devices that increase in resistance as they are bent or flexed (see Figs. 5.40 and 5.41). More commonly used for making virtual-reality data gloves to measure

■ **5.39** *Momentary contact switches*

■ **5.40** *Bend sensor*

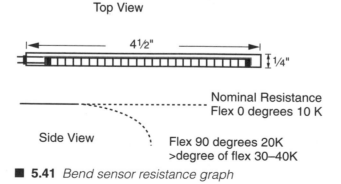

Top View

4½"

¼"

Nominal Resistance
Flex 0 degrees 10 K

Side View

Flex 90 degrees 20K
>degree of flex 30–40K

■ **5.41** *Bend sensor resistance graph*

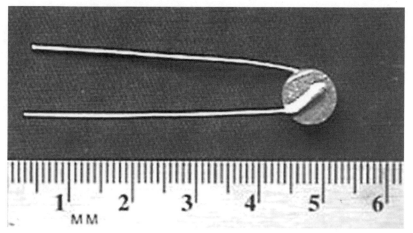

■ **5.42** *Thermistor*

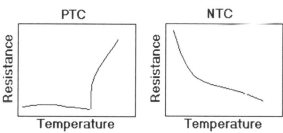

■ **5.43** *Positive (left) and negative (right) temperature co-efficient thermistor graphs*

the flexing of fingers, these versatile sensors can easily be adapted to robotics. The bend sensor makes an interesting feeler that can inform the robot of an obstacle.

I am reminded of a cat's whiskers. Cats use their whiskers to determine if a particular passageway is wide enough to pass through. If the whiskers on both sides of a cat's face touch each side of a passageway, the cat will most probably not try to pass through it. The bend sensors can be used in a similar manner.

### Heat

The most common heat sensor is the thermistor (see Fig. 5.42). This passive device changes resistance in proportion to its temperature. There are positive temperature coefficient and negative temperature coefficient thermistors (see Fig. 5.43). Thermal radiation can also be detected by piezoelectric materials as discussed earlier.

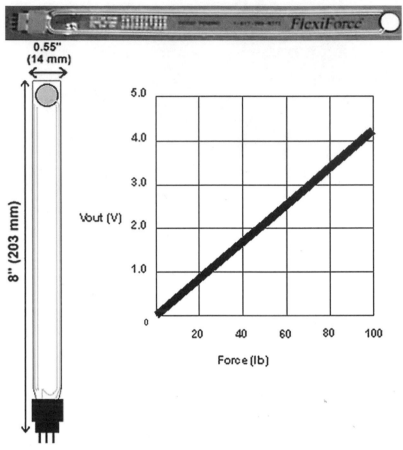

■ **5.44** *Flexiforce pressure sensor*

## Pressure sensor

Pressure sensors, shown in Fig. 5.44, are perfect for measuring forces. The "sensor" portion on the sensor is contained in the 14 mm × 14 mm pad at the end of the sensor. The resistance of the sensor decreases as force is applied. There are a variety of pressure ranges available from 0 to 1 pound (lb) up to 0 to 1000 lb.

## Smell

Currently no sensor exists that can approach the olfactory sense of the human nose. What is available are simple gas sensors that can detect toxic gases (see Fig. 5.45). The gas sensors can be used to create automatic (robotic) ventilation systems.

A simple sensor setup is shown in Fig. 5.46. The resistive element must be heated to become sensitive. The sensor incorporates its own heating unit, which is separately powered. The heater requires a regulated +5 V for proper operation and draws about 130 mA. The resistive element can be read like any other resistive sensor used thus far.

The potential for these gas sensors is greater than what is implied in the simple schematic. The gas sensors are not precise instruments. In other words, their response varies slightly from device to device. This "analog" property can be used to create a more sensitive smell detector.

Let's arrange eight sensors. The resistive element from each sensor is connected to an A/D convertor. A comparator circuit wouldn't do in this situation because precise and subtle variations in response are what we are looking for. To calibrate the device, a small amount of a known gas (smell) is released by the eight sensors. The response of each detector is measured by the A/D convertor and recorded by the main computer. Since the responses of the detectors will vary, an eight-number pattern is created for each smell.

Pattern matching is well established in neural networks. A neural network can be built using the information gathered that can not only measure but recognize different smells.

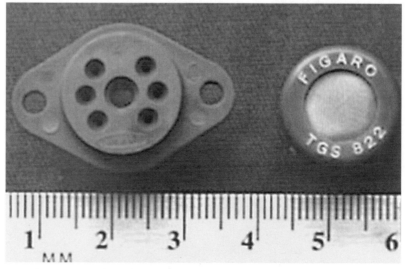

■ **5.45** *Toxic gas sensor*

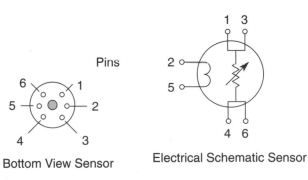

Pins

Bottom View Sensor

Electrical Schematic Sensor

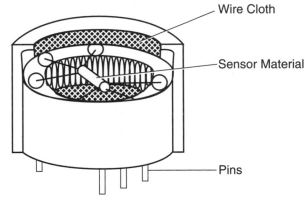

Wire Cloth

Sensor Material

Pins

Cutaway View — Gas Sensor

9V

7805

+5V

2

1  3

5

4  6

V

Gas Sensor

4.7K

■ **5.46** *Toxic gas sensor test circuit*

## Humidity

Passive resistive humidity sensors are a relatively new product that can be purchased.

## Testing sensors

When designing and building sensor systems, it's a good idea to test them before committing to use the system on a robot. One method that I used was to build a small mobile robot whose only function is to test sensors. That way reliability and response time can be determined before committing the sensors on a more elaborate robot.

The robot can test bump switches, light switches, bend sensors, and infrared and ultrasonic obstacle avoidance sensors. Other types of sensors may require a different test bed.

## Building a tester robot

Tester is the name I've given this small robot. The foundation of the robot is a small electric car that can be purchased for less than $10.00 (see Fig. 5.47).

The schematic for Tester is shown in Fig. 5.48. The sensor connects to the trigger input of a 555 timer set up as a monostable

■ **5.47** *Tester*

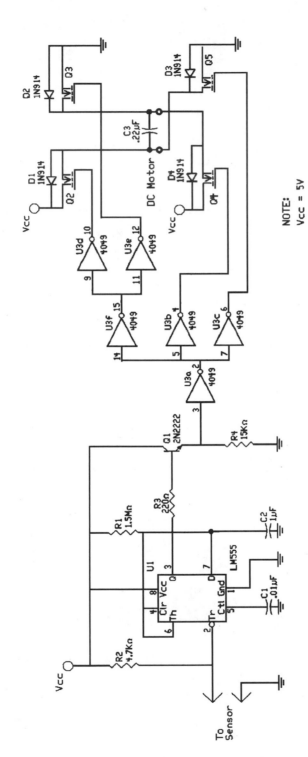

■ **5.48** *Tester circuit schematic*

pulse generator. The output (pin 3) of the 555 remains low until a negative pulse on the number 2 pin triggers its operation. Once triggered, the output (pin 3) of the 555 goes high for approximately 1 s.

The output of the 555 connects to a 2N2222 NPN transistor. An output is taken off the emitter of the transistor and connected to a buffer on the 4049 hex inverting buffer IC. The buffers on the 4049 chip are connected to a four-MOSFET (metal oxide semiconductor field-effect transistor) H-bridge that controls the drive motor.

When the output of the 555 timer is low, the H-bridge powers the robot's drive motor forward. The sensor to be tested is connected to the trigger input, pin 2 on the 555 timer. The sensor is wired in such a way as to cause a negative pulse (goes to ground) when it is activated or tripped. The negative pulse on pin 2 causes the output of the timer to go high for 1 s, which reverses the motor direction for 1 s.

Tester can be used to check a variety of sensors and transducers.

## Improving the tester robot

When I designed Tester, I had imagined most of the sensors I would test and use to be tiny miniature modules. This was not the case. In the process of prototyping different circuits I rarely had the time to produce a PCB, let alone miniaturize the circuit.

If I were to build another tester robot, I would use a much larger electric car as a foundation. Having a lot of room to work on the robot makes it easier to secure different types of sensors and circuits.

Parts for the projects outlined in this chapter are available from:

Images Company
P.O. Box 140742
Staten Island, NY 10314
(718) 698-8305
http://www.imagesco.com

# 6

# Intelligence

INTELLIGENCE PACKAGED IN A ROBOT TAKES ONE OF TWO forms: rule-based (expert) or neural. It's possible for both forms of intelligence to work in tandem. This synthesis of intelligence will be commonly used in robotics to create a robust intelligence system.

Expert (rule-based) intelligence programs are familiar to most people; these are programs written in high-level or low-level languages like C++, BASIC, and assembly. Neural systems on the other hand use electronic neurons and feedback to control (generate behavior of) the robot. This neural behavior-based robotic architecture was pioneered in the late 1940s and early 1950s by William Grey Walter. More recently, Rodney Brooks at the Massachusetts Institute of Technology (MIT) has been developing behavior-based robotic architecture under the name of "subsumption architecture." We will look at behavior-based robotics in Chap. 8.

In this chapter we will focus on rule-based systems and microcontrollers. Keep in mind that it is possible to mimic neural systems using rule-based systems programming. It is also noteworthy to know that almost all neural network software on the market today runs on existing rule-based computers, using rule-based programming that simulates neural networks.

## Microchip's PIC microcontroller

Adding intelligence in the form of a computer to a small robot or robotic system has never been easier. There are numerous single-chip computers (commonly know as microcontrollers) available that can do the job.

As the name implies, a *single-chip computer* is an entire computer system that lies within the confines of an integrated circuit (IC) chip. The microcontroller existing on the encapsulated sliver of silicon has features and similarities to our standard personal computer (PC). Primarily the microcontroller is capable of storing and running a program (most important feature). The microcontroller contains a central processing unit (CPU), random access memory (RAM), read only memory (ROM), input/output (I/O) lines, serial and parallel ports, timers, and sometimes other built-in peripherals like analog-to-digital (A/D) and digital-to-analog (D/A) converters.

## Why use a microcontroller?

The microcontroller's ability to store and run unique programs makes it extremely versatile. For instance, one can program a microcontroller to make decisions (perform functions) based on predetermined situations (I/O line logic) and sensor readings. Its ability to perform math and logic functions allows it to mimic sophisticated logic and electronic circuits. Still other programs can make the microcontroller behave like a neural or fuzzy logic controller.

The output of the microcontroller can control direct current (DC) motor drives [using DC or pulse-width modulation (PWM)], servo motor positioning, stepper motors, etc. Programming a robot's microcontroller to respond to sensor readings or a communication link creates an intelligent, responsive robot. Microcontrollers are responsible for the "intelligence" in most smart devices on the consumer market and will be the intelligence in our robots.

## PIC programming overview

Programming PIC microcontrollers is a three-step process. Before you can program, however, you need to purchase two items, the PICBASIC compiler program and the EPIC programmer (a programming carrier board). These two items do not include the PIC microcontroller chip or its support components. I recommend beginning with the 16F84 PIC microcontroller because it is a versatile 18-pin chip with 13 I/O lines and rewritable flash memory. This flash memory allows you to reprogram the PIC microcontroller chip 1000 times. This is really useful when testing and debugging your programs and circuits.

The PICBASIC compiler (see Fig. 6.1) runs on a standard PC. The program may be run in DOS or in an "MS-DOS Prompt" window in the Windows environment. Hereafter the MS-DOS Prompt window will be referred to simply as a DOS window. The DOS program will

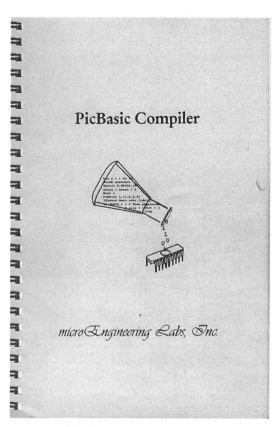

**■ 6.1** *PICBASIC compiler*

run on everything from an XT class PC running DOS 3.3 and higher. The compiler supports a large variety of PIC microcontrollers. The compiler generates machine language (ML) hex code that may be used with other programming carrier boards. The cost for PICBASIC compiler software is $99.95.

The EPIC programming carrier board (see Fig. 6.2) has a socket for inserting the PIC chip and connecting it to the computer, via the printer port, for programming. The programming board connects to the computer's printer port (also called the parallel port) using a DB25 cable. If the computer only has one printer port with a printer connected to it, the printer must be temporarily disconnected when programming PIC chips. As with the PICBASIC Compiler, the EPIC programming carrier board supports a large variety of PIC microcontrollers. The cost for the EPIC programming board with the EPIC programming diskette is $59.00.

The PIC 16F84 is shown in Fig. 6.3. It is a versatile microcontroller with flash memory. Flash memory as stated before is rewritable

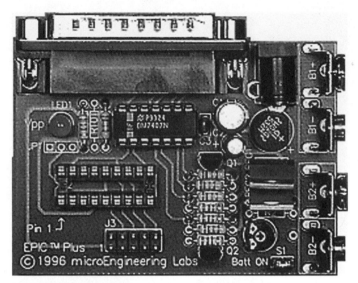

■ **6.2** *EPIC programming board*

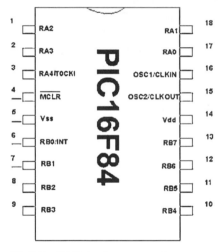

| | PIC16F84 | |
|---|---|---|
| 1 | RA2 | RA1 18 |
| 2 | RA3 | RA0 17 |
| 3 | RA4/T0CKI | OSC1/CLKIN 16 |
| 4 | MCLR | OSC2/CLKOUT 15 |
| 5 | Vss | Vdd 14 |
| 6 | RB0/INT | RB7 13 |
| 7 | RB1 | RB6 12 |
| 8 | RB2 | RB5 11 |
| 9 | RB3 | RB4 10 |

■ **6.3** *16F84 microcontroller*

## Features

**GENERAL**

RISC CPU 35 single-word instructions
Operating speed DC, 10-MHz clock input
1K program memory
14-bit-wide instructions
8-bit-wide data path
Direct, indirect and relative addressing
1000 erase/write cycles

**PERIPHERAL**

13 I/O pins with individual direction control
High current sink/source for direct LED drive
- 25 mA sink max per pin
- 20 mA source max per pin
TMR0: 8-bit timer/counter with 8-bit
 programmable prescaler

memory. The onboard flash memory can endure a minimum of 1000 erase-write cycles. So you can reprogram and reuse the PIC chip at least 1000 times. The program retention time, if you decide not to rewrite the program, is approximately 40 years. The 18-pin 16F84 chip devotes 13 of its pins to I/O. Each pin may be independently programmed as an input or output. The pin's status (I/O direction control) may also be changed on the fly via programming. Other features include power on reset, power-saving sleep mode, power-up timer, and code protection. Additional features and architecture details of the PIC 16F84 will be given as we continue.

## Software installation

Install the compiler (PICBASIC) and programming (EPIC) software according to the directions provided in their manuals. I created a directory on my computer's hard drive called APPLICS. I used the DOS path command so that I could run both the compiler and programming software from this directory. I wrote and saved all my programming text files in the APPLICS directory also. For complete software installation directions with a basic DOS commands tutorial, along with numerous PIC microcontroller applications, read my *PIC Microcontroller Project Book* (McGraw-Hill, New York, 2000).

## Step 1: Writing the BASIC language program

PICBASIC programs are written using any word processor that is able to save its text file as ASCII or DOS text. Every word processor I have worked with has this option. Use the Save As command and choose MS-DOS text, DOS text, or ASCII text. The finished text file is compiled into a program by the PICBASIC compiler. If you don't own a word processor, you can use Windows Notepad, which is included with Windows 3.X, 95, and 98, to write the BASIC language source file. (In Windows, look under Accessories.) At the DOS level you can use the Edit program to write text files.

When you save the file, save it with a .bas suffix. So if you were saving a program named Wink, you would save it as wink.bas.

## Step 2: Using the compiler

The PICBASIC compiler program is started by entering the command pbc followed by the name of the text file. For example, if the text file we created is named wink.bas, then at the DOS command prompt enter

pbc wink.bas

The BASIC compiler compiles the text file and creates two additional files, an .asm (assembly language) file and a .hex (hexadecimal) file.

The wink.asm file is the assembly language equivalent to the BASIC program. The wink.hex file is the machine code of the program written in hexadecimal numbers. It is the .hex file that is loaded into the PIC chip.

If the compiler encounters errors when compiling the BASIC source code, it will list each error it has found along with the line number where the error is located and terminate. The errors listed need to be corrected in the BASIC source code before it will successfully compile.

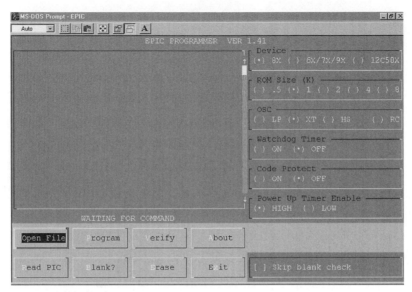

■ **6.4** *EPIC programming screen*

## Step 3: Programming the PIC chip

Connect the EPIC programming board to the computer's printer port using a DB25 cable. Start the DOS programming software. At a DOS command prompt, enter

```
EPIC
```

Figure 6.4 is a picture of the programming screen. Use the Open File option and select wink.hex from the files displayed in the dialog box. The file will load, and numbers will be displayed in the window on the left. Insert the 16F84 into the socket, and then press the program button. The PIC microcontroller is programmed and ready to go to work.

## First BASIC program

We are ready to write our first program. Enter this program in your word processor exactly as it is written:

```
'First BASIC program to wink two LEDs connected to port B.
Loop: High 0    ' Turn on LED connected to pin RB0
      Low 1     ' Turn off LED connected to pin RB1
      Pause 500    ' Delay for 0.5 seconds
      Low 0     ' Turn off LED connected to pin RB0
      High 1    ' Turn on LED connected to pin RB1
      Pause 500    ' Delay for 0.5 seconds
      Goto loop    ' Go back to loop and blink and wink LEDs
      End
```

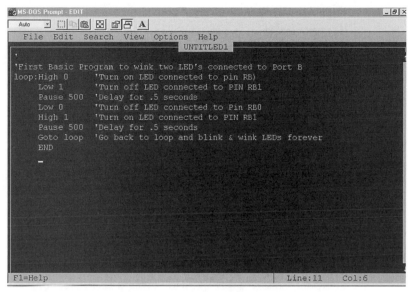

**■ 6.5** *PICBASIC program text file*

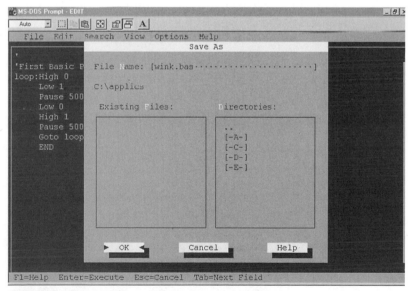

**■ 6.6** *Saving text file*

See Fig. 6.5. Save the above as a text file using the Save function under the File menu. Name the file wink.bas (see Fig. 6.6). If by accident you saved the file as wink.txt, don't get discouraged. You can do a Save As from the Edit program (under the File menu) and rename the file wink.bas.

## Compile

The PICBASIC compiler must be run from DOS or from a DOS prompt window within Windows. I ran the PICBASIC compiler from the APPLICS directory. Make sure the wink.bas file is also in the PICBASIC directory. The PICBASIC compiler is compatible with dozens of different PIC microcontrollers. In order to compile a program for a specific microcontroller, the PICBASIC compiler needs to know which microcontroller we are using. To compile a program for the 16F84, we add `-p16F84` to our `pbc` command.

So the complete command is `pbc -p16f84 wink.bas`. At the DOS prompt, type in the command and hit the Enter key (see Fig. 6.7).

```
C:/APPLICS>pbc -p16F84 wink.bas
```

The compiler displays an initialization copyright message and begins processing the BASIC source code (see Fig. 6.8). If the BASIC source code is without errors, it will create two additional files. If the compiler finds any errors, it displays a list of errors with their line numbers. Match the line numbers in the error message to the line numbers in the .bas text file to locate where the errors occurred. The errors need to be corrected before the compiler can compile the source code correctly.

You can look at the files by using the `dir` directory command. Type `dir` at the command prompt and hit Enter (see Fig. 6.9).

```
C:\APPLICS> dir
```

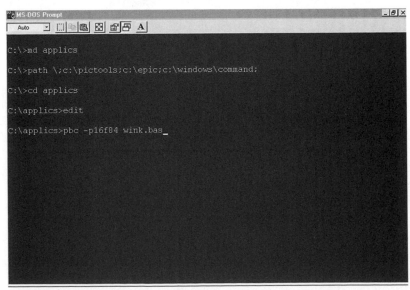

■ **6.7** *Entering compile command*

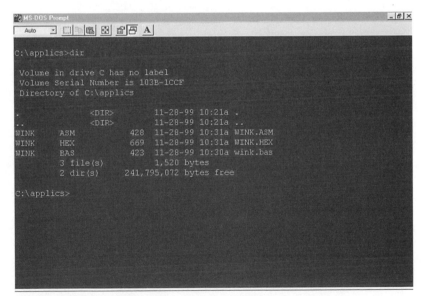

■ **6.8** *Compiler compiling program*

■ **6.9** *Directory command*

The di r command displays all the files and subdirectories within the subdirectory where it is issued. In Fig. 6.9, we can see the two additional files the compiler created. One file is the wink.asm file and is the assembler source code file that automatically initiated the macroassembler to compile the assembly code to machine language hex code. The hex code file is the second file created called wink.hex.

■ **6.10** *EPIC programming board*

## Programming the PIC chip

To program the PIC chip, we connect the EPIC programming carrier board (see Fig. 6.10) to the computer. The EPIC board connects to the printer port. If your computer has only one printer port, disconnect the printer, if one is connected, and attach the EPIC programming board using a 6-foot (ft) DB25 cable.

When connecting the programming board to the computer, make sure there isn't a PIC microcontroller installed in the board. If you have an alternating current (AC) adapter for the EPIC programming board, plug it into the board. If you do not have the AC adapter, attach two fresh 9-volt (9V) batteries and connect the "Batt ON" jumper to apply power. The programming board must be connected to the printer port with power applied to the board before running the software. If not, the software will not see the programming board connected to the printer port and will give the error message "EPIC programmer not connected."

When power is applied and it is connected to the printer port, the light-emitting diode (LED) on the EPIC programmer board may be on or off at this point. Do not insert a PIC microcontroller into the programming board socket until the EPIC programming software is running.

## The EPIC programming board software

There are two versions of the EPIC software: EPIC.exe for DOS and EPICWIN.exe for Windows. The Windows software is 32-bit. It may be used with Windows 95, 98, and NT, but not 3.X.

### Using the EPIC DOS version

If using Windows 95 or higher, you could either open an MS-DOS prompt window or restart the computer in the DOS mode. Windows 3.XX users should end the Windows session.

Assume we are still in the same DOS session and we have just run the pbc compiler on the wink.bas program. Copy the wink.hex file into the EPIC subdirectory. At the DOS prompt, type "EPIC" and hit Enter to run the DOS version of the EPIC software (see Fig. 6.11).

EPIC's opening screen is shown in Fig. 6.12. Use the mouse to click on the Open button or press Alt-O on your keyboard. Select the wink.hex file (see Fig. 6.13). When the hex file loads, you will see a list of numbers in the window on the left (see Fig. 6.14). This is the machine code of your program. On the right-hand side of the screen are configuration switches that we need to check before we program the PIC chip.

Let's go through the configuration switches one by one.

- ☐ Device: Sets the device type. Set it for 8X.
- ☐ ROM size (K): Sets memory size. Choose 1.
- ☐ OSC: Sets oscillator type. Choose XT for crystal.
- ☐ Watchdog timer: Choose On.
- ☐ Code protect: Choose Off.
- ☐ Power-up timer enable: Choose High.

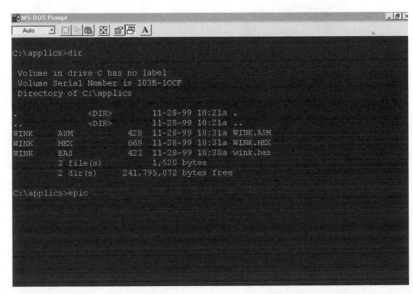

■ **6.11** *EPIC command*

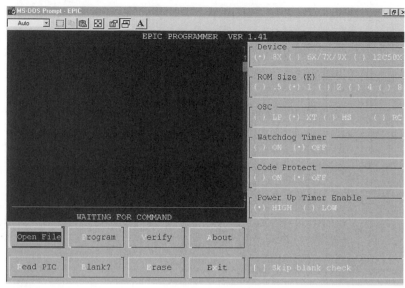

■ **6.12** *EPIC programming screen*

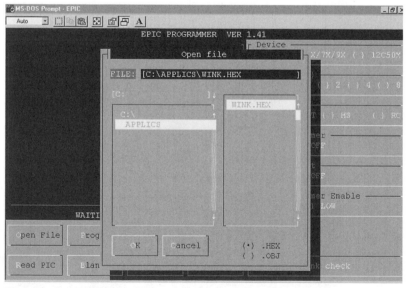

■ **6.13** *Selecting hex file*

After the configuration switches are set, insert the PIC 16F84 micro-controller into the socket. Click on Program or press Alt-P on the keyboard to begin programming. The EPIC program first looks at the microcontroller chip to see if it is blank. If the chip is blank, the EPIC program installs your program into the microcontroller. If

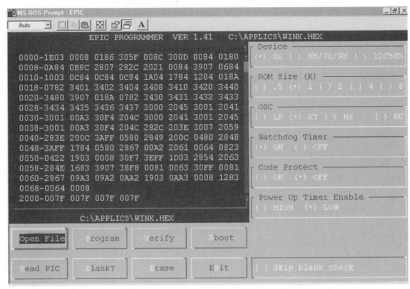

■ **6.14** *Hex file loaded into EPIC program*

the microcontroller is not blank, you are given the options to cancel the operation or overwrite the existing program with the new program. If there is an existing program in the PIC chip's memory, write over it. The machine language code lines are highlighted as the PIC is programmed. When it is finished, the microcontroller is programmed and ready to run.

### Testing the PIC microcontroller

The schematic shows how minimal is the number of components needed to get your microcontroller up and running. Primarily you need a pull-up resistor on pin 4 (MCLR), a 4-megahertz (MHz) crystal with two [22-picofarad (pF)] capacitors, and a 5V power supply.

The two LEDs and the two current-limiting resistors connected in series with the LEDs are the output. It allows us to see that the microcontroller and program are functioning. Assemble the components as shown in the schematic of Fig. 6.15 on the solderless breadboard. When you are finished, your work should appear as in Fig. 6.16.

While the specifications sheet on the 16F84 states the microcontroller will operate on voltages from 2V to 6V, I provided a regulated 5V power supply for the circuit. The regulated power supply consists of a 7805 voltage regulator and two filter capacitors.

\* Capacitors connected to crystals are 22pF

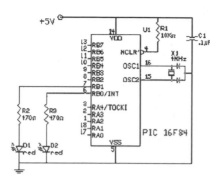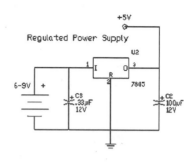

■ **6.15** *Schematic*

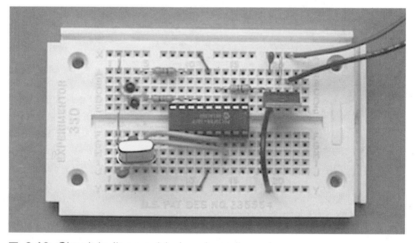

■ **6.16** *Circuit built on solderless breadboard*

## Wink

Apply power to the circuit. The LEDs connected to the chip will alternately turn on and off. Wink...wink.... Now you know how easy it is to program these microcontrollers and get them up and running.

As you gain experience, using the compiler and programmer will become second nature. You won't even consider them as steps anymore. The real challenge will be in writing the best PICBASIC programs possible. And that is as it should be.

## Troubleshooting the circuit

There is not too much that can go wrong here. If the LEDs don't light up, the first thing I would check is the orientation of the LEDs. If they are put in backwards, they will not light.

## PICBASIC Pro compiler

There is also a high-end version of the PICBASIC compiler named the PICBASIC Professional Compiler. The Pro version compiler is considerably more expensive, retailing for $249.95. The Pro version has a greater number of and much richer BASIC commands than are in the standard compiler package. A few of the additional commands to be found in the Pro version allow the use of Interrupts, direct control of light-crystal display (LCD) modules, dual-tone multifrequency (DTMF) out, and X-10 commands, to name a few.

While a more sophisticated package, the compiler does not handle two of my favorite (and very powerful) BASIC commands: Peek and Poke. While the commands are listed "as functional" in the Pro manual, it is emphasized that "Peek and Poke should never be used in a PICBASIC Pro program." This is unfortunate, personal feelings aside, because it destroys upward compatibility of any PICBASIC programs that use the Peek and Poke commands.

# New IDE features

Recently, both the PICBASIC and PICBASIC Pro compilers are being packaged with an additional diskette that contains a Windows integrated development environment (IDE) interface called CodeDesigner Lite (see Fig. 6.17). CodeDesigner Lite allows one

■ **6.17** *CodeDesigner Lite*

to write and compile PICBASIC code in a Windows environment. Each statement is color-coded, making it much easier to spot errors and read through your code. The freebie version allows you to write programs up to 150 lines and open up three source files at once for easy copy and paste.

The most important feature of the CodeDesigner IDE interface is that it allows you to first write the program, then compile the program into a hex file, and finally (in theory) program the microcontroller while in the same window. This reduces program development time. Typically, I write while in DOS or an MS-DOS prompt window. I write the program text file using the DOS Edit program. When finished, I then exit Edit and manually compile the program. If there is a problem (more times than not), I then restart Edit and debug the code. When the program is completely debugged, I load the program into the PIC microcontroller using the EPIC software and programming board. At this point the microcontroller/circuit is tested. If it functions properly, I'm finished; if not, I begin rewriting the program.

In using CodeDesigner, the ease with which you can write and debug PICBASIC programs and load them into the microcontroller increases productivity. My experience is that I can code and debug my programs while in Windows, but to program a microcontroller, I still must drop down into DOS.

While the freebie version (CodeDesigner Lite) is functional, if you like it, you can then upgrade to the full-featured CodeDesigner. CodeDesigner is available in a hobbyist version for $45.00 and a standard version for $75.00.

The hobbyist version of CodeDesigner only works with the PICBASIC compiler. The standard version will work with both the PICBASIC and PICBASIC Pro compilers. Some of the advanced features of CodeDesigner include

☐ AutoCodeCompletion: CodeDesigner makes writing code much easier with smart pop-up list boxes that can automatically fill in statements and parameters for you.

☐ Multiple document support.

☐ Line error highlighting: CodeDesigner will read error data and highlight error lines when you compile your PICBASIC project.

☐ QuickSyntaxHelp: The QuickSyntaxHelp feature displays statement syntax when you type in a valid PICBASIC statement.

☐ Statement description: Statement descriptions are displayed in the status bar when you type in a valid PICBASIC statement.

- ☐ Statement help: Simply position your cursor over a PICBASIC statement and get statement-specific help.
- ☐ Label listbox: The label listbox displays the current label and allows you to select a label from the list to jump to the selected label.
- ☐ Colored PICBASIC Syntax: Set colors for reserved words, strings, numbers, comments, defines, etc. Colored PICBASIC syntax makes for easy code reading.
- ☐ Bookmarks: Never lose your place again. CodeDesigner allows you to set bookmarks.
- ☐ Multiple undo/redo: Didn't want to delete that last line? No problem. Simply click on the Undo button.
- ☐ Multiple views: Multiple views of your source code allow you to easily edit your code.
- ☐ Print source code.
- ☐ Drag and drop text.
- ☐ Row/column-based insert, delete, and copy.
- ☐ Search and replace.
- ☐ Compile and launch device programmer.

### Software installation

When it is being installed, CodeDesigner creates a subdirectory in the Program Files directory and installs itself there. It puts a CodeDesigner shortcut on the Start, Program menu in Windows.

## First PICBASIC Pro program

This program is identical in function (not code) to the wink.bas PICBASIC program. Start CodeDesigner (Lite) (see Fig. 6.18) and enter the following code:

```
'Wink program
'Blinks and winks two LEDs connected to port B
Loop:
High PORTB.0    ' Turn on LED connected to RB0
Low PORTB.1     ' Turn off LED connected to RB1
Pause 500    ' Wait 1/2 second
Low PORTB.0     ' Turn off LED connected to RB0
High PORTB.1    ' Turn on LED connected to RB1
Pause 500    ' Wait 1/2 second
GoTo Loop    ' Loop back—repeat cycle blink and wink forever
```

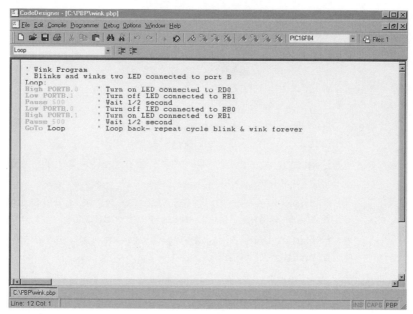

```
' Wink Program
' Blinks and winks two LED connected to port B
Loop:
High PORTB.0          ' Turn on LED connected to RD0
Low PORTB.1           ' Turn off LED connected to RB1
Pause 500             ' Wait 1/2 second
Low PORTB.0           ' Turn off LED connected to RB0
High PORTB.1          ' Turn on LED connected to RB1
Pause 500             ' Wait 1/2 second
GoTo Loop             ' Loop back- repeat cycle blink & wink forever
```

■ **6.18** *PICBASIC Pro program written in CodeDesigner*

CodeDesigner defaults to writing code for the PIC 16F84 micro-controller. This is the microcontroller I recommend you start with. To change the device, simply pull down the device menu and select the appropriate microcontroller.

To compile the program, either select compile under the Compile menu or hit F5. CodeDesigner automatically starts the PICBASIC Pro compiler to compile the program. Before you attempt to compile a program, set up the "compiler options" under the Compile menu. CodeDesigner needs you to tell it where (which directory) to find the PICBASIC Pro program, and where to save the compiled and source files.

Once the program is compiled, we can go to the next step of loading the program into a PIC microcontroller chip using the EPIC programmer. Follow the instructions for the EPIC program described previously for the PICBASIC compiler.

### The EPIC programmer and CodeDesigner

If you prefer, you could also program the chip from CodeDesigner. Select "launch programmer" from the Programmer menu, or hit F6. CodeDesigner automatically starts the EPICWIN.exe Windows software.

With the EPIC Windows software started, set the configuration switches one by one under the Options menu.

☐ Device: Sets the device type. Set it for 16F84 (default).

☐ Memory size (K): Sets memory size. Choose 1.

☐ OSC: Sets oscillator type. Choose XT for crystal.

☐ Watchdog timer: Choose On.

☐ Code protect: Choose Off.

☐ Power-up timer enable: Choose High.

After the configuration switches are set, insert the PIC 16F84 microcontroller into the open socket on the EPIC programming board. If you receive an error message "EPIC programmer not found" when CodeDesigner starts the EPIC Windows program (see Fig. 6.19), you have the option of either troubleshooting the problem or using the EPIC DOS program. For instructions on using EPIC software (DOS version), see the PICBASIC compiler section. The schematic for the circuit is the same schematic used for the PICBASIC compiler.

### Wink

Apply power to the circuit. The LEDs connected to the PIC microcontroller will alternately turn on and off.

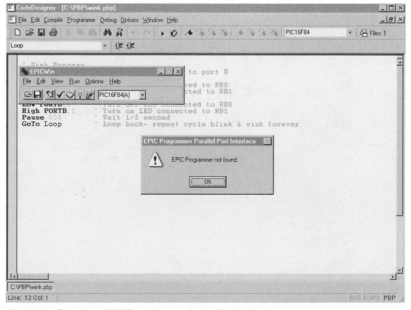

■ **6.19** *Starting EPIC software from CodeDesigner*

## Moving forward—applications

It's now time to show you how to put these microcontrollers to work. You have the basic knowledge needed to program the 16F84 microcontroller. The remainder of this chapter illustrates how to perform basic electrical functions using the microcontroller. These functions are ubiquitous in microcontroller use in electronic circuits and designs.

To begin, let's examine how a microcontroller can detect simple switch closure. The microcontroller can detect transistor-transistor logic (TTL) levels on any of its 13 I/O pins. We use these logic levels in conjunction with switches (see Fig. 6.20) for closure detection.

### Reading switches—logic low

In Fig. 6.20 the switch labeled A keeps the I/O pin at a logic high until the switch is closed. Once closed, the I/O pin is brought to ground, or a logic low. When the microcontroller senses switch closure, it can perform any number of operations or control functions. In our example it will blink an LED. Keep in mind that the LED may represent a transistor, transducer, electronic circuit, or another microcontroller/computer.

The program for the PICBASIC compiler is as follows:

```
'PICBASIC Compiler
'REM test switch low
'Initialize variables
input 4    'Set pin RB4 to read switch
start:
if pin4 = 0 then blink    'If switch is low, then blink LED
goto start    'If not, check switch again
blink:    'Blink routine
```

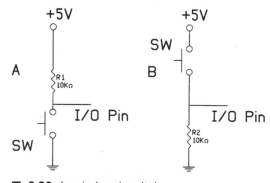

■ **6.20** *Logic level switches*

```
high 0    'Bring RB0 high to light LED
pause 250    'Wait 1/4 second
low 0    'Bring RB0 low to turn off LED
pause 250    'Wait 1/4 second
goto start    'Check switch again
```

The program for the PICBASIC Pro compiler is as follows:

```
'REM PICBASIC Compiler Pro
'Rem test switch low
input portb.4    'Set pin RB4 to read switch
start:
if portb.4 = 0 then blink    'If switch is low, then blink LED
goto start    'If not, check again
blink:    'Blink LED routine
high 0    'Bring RB0 high to light LED
pause 250    'Wait 1/4 second
low 0    'Bring RB0 low to turn off LED
pause 250    'Wait 1/4 second
goto start    'Check switch again
```

The schematic for the read-switch-low circuit is shown in Fig. 6.21. The switch is connected to an I/O pin labeled RB4. The LED is connected to RB0 through a 470-ohm current-limiting resistor.

### Reading switches—logic high

These programs and schematic are the complement to the previous examples. Look back to Fig. 6.20, example B. The switch labeled B keeps the I/O pin at a logic low level. When the switch is closed, the I/O pin is brought to a logic high level.

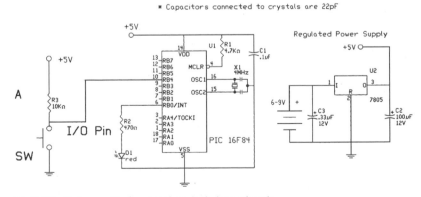

■ **6.21** *Schematic for read-switch-low circuit*

The program for the PICBASIC compiler is as follows:

```
'PICBASIC Compiler
'REM test switch high
input 4     'Set pin RB4 to read switch
start:
if pin4 = 1 then blink    'If switch is high, then blink LED
goto start    'If not, check switch again
blink:    'Blink routine
high 0    'Bring RB0 high to light LED
pause 250    'Wait 1/4 second
low 0    'Bring RB0 low to turn off LED
pause 250    'Wait 1/4 second
goto start    'Check switch again
```

The program for the PICBASIC Pro compiler is as follows:

```
'PICBASIC Compiler Pro
'REM test switch high
input portb.4    'Set pin RB4 to read switch
start:
if portb.4 = 1 then blink    'If switch is high, then blink LED
goto start    'If not, check again
blink:    'Blink LED routine
high 0    'Bring RB0 high to light LED
pause 250    'Wait 1/4 second
low 0    'Bring RB0 low to turn off LED
pause 250    'Wait 1/4 second
goto start    'Check switch again
```

The schematic for the read-switch-high circuit is shown in Fig. 6.22. The switch is connected to the I/O pin labeled RB4. The LED is connected to RB0 through a 470-ohm current-limiting resistor.

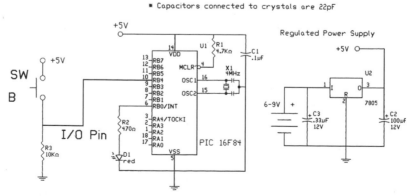

■ **6.22** *Schematic for read-switch-high circuit*

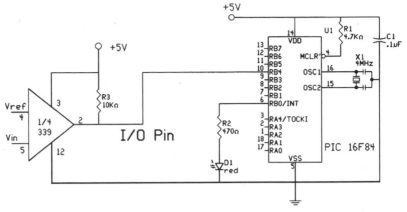

* Capacitors connected to crystals are 22pF

■ **6.23** *Schematic for read comparator*

## Reading comparators

The microcontroller can also read logic levels from other micro-controllers, circuits, or ICs. As an example, look at Fig. 6.23. In this schematic the microcontroller is set to read the output of an comparator. Since the output of an LM339 comparator is equivalent to an open collector of an NPN transistor, it is usually brought high by using an external pull-up resistor. The comparator is read by the microcontroller using the same programs that detect a logic low.

## Reading resistive sensors

The PIC microcontroller is able to read resistive sensors that vary in resistance from 5K to 50K ohms directly. The types of resistive sensors one can connect to the microcontroller are numerous, for instance, photoresistors [cadmium sulfide (CdS) cells], thermistors (PTC and NTC types), toxic gas sensors, bend sensors, and humidity sensors. The microcontroller reads the resistance by timing the discharge of a capacitor through the resistive device (see Fig. 6.24).

The command to read a resistive sensor is

```
Pot pin, scale, var
```

where Pot is the command, and pin is the pin number the resistive sensor is connected to. Variable scale is used to adjust the RC constant. For a large RC constant, scale should be set low, and for a small RC constant, scale should be set to its maximum value of 255. When the value of scale is set correctly, the value contained in the var variable will be near zero at minimum resistance value and to 255 near maximum resistance value.

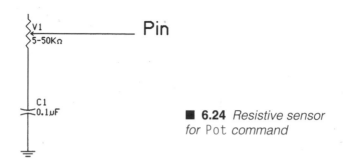

**■ 6.24** *Resistive sensor for* Pot *command*

* Capacitors connected to crystals are 22pF

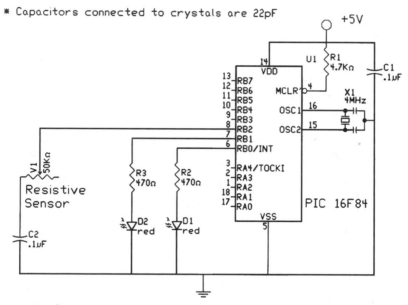

**■ 6.25** *Schematic for* Pot *command*

The value of scale needs to be determined experimentally. To find a good scale value, set the resistive device under measurement to its maximum resistance and read the var variable with scale set to 255. Under these conditions, the value held in the var variable will contain a reasonable value for scale.

A schematic of a basic circuit is shown in Fig. 6.25. For the resistive sensor you can connect a 50K-ohm potentiometer. As the potentiometer is varied, one of the LEDs will be lit depending upon the value held in the variable B0. If the resistance value read is above 125, LED 1 will light; if not, LED 2 will light.

The program for the PICBASIC compiler is as follows:

```
' PICBASIC Compiler ** reading resistance type sensors **
'Photoresistor test program
```

```
' Set Up
start:
pot 2,255, b0      'Read sensor on RB2
if b0 > 125 then l1     'If more than 100, light LED 1
if b0 <= 125 then l2    'If less than 100, light LED 2
l1:    'Light LED 1 routine
high 0    'Light LED 1
low 1     'Turn off LED 2
goto start    'Repeat
l2:    'Light LED 2 routine
high 1    'Light LED 2
low 0     'Turn off LED 1
goto start    'Repeat
```

The program for the PICBASIC Pro compiler is as follows:

```
' PICBASIC Pro Compiler ** reading resistance type sensors **
'Photoresistor test program
' Set Up
output portb.0    'Set RB0 as output
output portb.1    'Set RB1 as output
b0 var byte
start:
pot portb.2,255, b0    'Read sensor on RB2
if b0 > 125 then l1     'If more than 100, light LED 1
if b0 <= 125 then l2    'If less than 100, light LED 2
l1:    'Light LED 1 routine
high portb.0    'Light LED 1
low portb.1     'Turn off LED 2
goto start    'Repeat
l2:    'Light LED 2 Routine
high portb.1    'Light LED 2
low portb.0     'Turn off LED 1
goto start    'Repeat
```

One can make the demonstration a little more interesting by substituting a CdS photoresistive cell in place of the potentiometer in the circuit. If the proper CdS cell is chosen, for instance, one with a dark resistance around 50K to 100K ohms and with a light saturation resistance of 10K ohms or less, LED 1 will be lit when the photoresistor is covered or in darkness. In bright light, LED 2 will be lit.

It's possible to read the numerical value of the pot variable by serially sending the variable to a serially interfaced LCD display or

RS232 computer connection. The command to send the information out serially is

```
Serout Pin, Mode, Var
```

While we are not doing serial communication right now, it's important that you know you can.

## Servo motors

Servo motors are geared direct current (DC) motors with a positional feedback control that allows the rotor of the motor to be accurately positioned. The shaft of most hobbyist servo motors can be positioned through a minimum of 90 degrees of rotation (±45 degrees). There are three wires to the servo motor. Two leads are for power, typically 4.5V to 6V, and ground. The third wire feeds the position control signal to the servo motor. The position control signal is a variable-width pulse. The pulse is varied between 1 and 2 milliseconds (ms). The width of the pulse controls the position of the servo motor's shaft.

Controlling servo motors with a PIC microcontroller is easy. The 1- to 2-ms control pulse signal must be sent to the motor 50 to 60 times a second.

The `pulsout` command generates a pulse on the pin specified, for the period specified [in 10-microsecond (μs) increments]. So the command `pulsout 1, 150` will place a 1.5 ms (10 μs × 150 = 1500 or 1.5 ms) pulse on pin 1. The 1.5-ms pulse will position the servo motor's shaft at midposition.

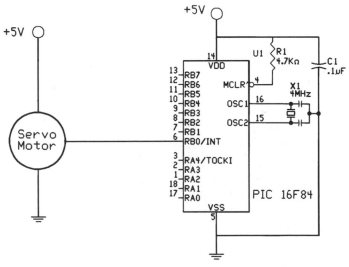

■ **6.26** *Schematic for servo motor*

### Servo sweep program

The demonstration program will sweep the servo rotator left to right and back again like a radar dish antenna. The schematic is shown in Fig. 6.26. Here is the PICBASIC compiler program:

```
'Servo motor sweep program
'PICBASIC Compiler
'Programs sweeps left to right and back again
b0 = 100    'Initialize at left position
sweep:    'Sweep routine
pulsout 0,b0    'Send pulse to servo motor
pause 18    'Wait 18 ms (50 to 60 Hz)
b0 = b0 + 1    'Increment pulse width
if b0 > 200 then sweepback    'End of sweep?
goto sweep    'No, continue sweeping
sweepback:    'Sweepback routine
b0 = b0 - 1    'Decrement pulse width
pulsout 0, b0    'Send pulse to servo motor
pause 18    'Delay to send 50 to 60 Hz
if b0 < 100 then sweep    'End of sweepback
goto sweepback    'No
```

The PICBASIC Pro compiler program is as follows:

```
'Servo motor sweep program
'PICBASIC Pro Compiler
'Programs sweeps left to right and back again
b0 var byte
b0 = 100    'Initialize at left position
sweep:    'Sweep routine
pulsout portb.0,b0    'Send pulse to servo motor
pause 18    'Wait 18 ms (50 to 60 Hz)
b0 = b0 + 1    'Increment pulse width
if b0 > 200 then sweepback    'End of sweep?
goto sweep    'No, continue sweeping
sweepback:    'Sweepback routine
b0 = b0 - 1    'Decrement pulse width
pulsout portb.0, b0    'Send pulse to servo motor
pause 18    'Delay to send 50 to 60 Hz
if b0 < 100 then sweep    'End of sweepback
goto sweepback    'No
```

## Fuzzy logic and neural sensors

We are presented with a few interesting possibilities regarding the interpretation of sensor readings. We can have the microcontroller mimic the function of neural and/or fuzzy logic devices.

## Fuzzy logic

In 1965, Lotfi Zadah, a Professor at the University of California at Berkeley, first published a paper on fuzzy logic. Since its inception, fuzzy logic has been both hyped and criticized.

In essence, fuzzy logic attempts to mimic in computers the way people apply logic in grouping and feature determination. A few examples should clear this "fuzzy" definition. For instance, how is a warm, sunny day determined not to be warm but to be hot instead, and by whom? The threshold of when someone considers a warm day hot depends on a person's personal heat threshold and the influence of his or her environment (see Fig. 6.27).

There is no universal thermometer that states at 81.9 degrees Fahrenheit (°F) it is warm and at 82°F it is hot. Extending this example further, a group of people living in Alaska has a different set of temperature values for hot days when compared to a group of people living in New York, and both these values will be different from that of a group of people living in Florida. And let's not forget seasonal variations. A hot day is a different temperature in winter than summer. So what this boils down to is that classifications (for example, "What is a hot day?") may be a range of temperatures determined by the opinions of a group of people. Further classifications can be differentiated by different groups of people.

Any particular temperature will find membership in the group where it fits into the range of values. Sometimes a temperature will fit into two overlapping groups. True membership will then be determined by how a particular temperature varies from the median values.

The idea of group and range classifications can be applied to many other things, like navigation, speed, and height. Let's use height for

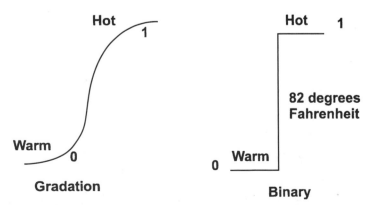

■ **6.27** *Grading temperature from warm to hot, gradually or by step*

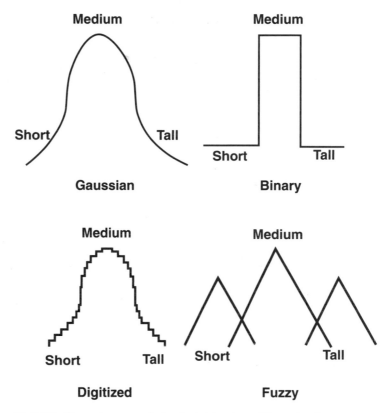

■ **6.28** *Grouping people by height using different schemes*

one more example. If we graph the height of 1000 people, our graph will resemble the first graph shown in Fig. 6.28. We can use this graph of heights to classify the heights into groups: short, medium, and tall. If we applied a hard rule that stated everyone under 5′ 7″ is short and everyone taller then 6′ 0″ is tall, our graph will resemble the second graph. This classifies a person who is 5′ 11.5″ tall as "medium," when in actuality the person's height is closer to the tall (6′ 0″ and over) group.

Instead of hard-and-fast rules typically employed by computers, people typically use soft and imprecise logic, or fuzzy logic. To implement fuzzy logic in computers we define groups and quantify the membership in that group. Groups overlap, as seen in the fourth graph of Fig. 6.28. So the person who is 5′ 11.5″ tall is almost out of the medium group (small membership) and well into the tall group (large membership).

Fuzzy logic provides an alternative to the digitized graph shown as the third graph of Fig. 6.28. A high-resolution digitized graph is

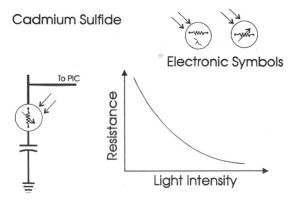

Cadmium Sulfide

Electronic Symbols

Resistance

Light Intensity

■ **6.29** *Electrical function of a CdS photoresistive cell*

also accurate in classifying height. Why would one choose a fuzzy logic method over a digitized model function? The fuzzy logic method has simplified mathematics and learning functions.

To implement fuzzy logic in a PIC microcontroller, one assigns a numeric range to a group. This is what we will do in our next project.

### Building a fuzzy logic light tracker

The project we will build now is a fuzzy logic light tracker. The tracker follows a light source using fuzzy logic.

The sensors needed for the tracker are two CdS photocells. These photocells are light-sensitive resistors (see Fig. 6.29). The resistance varies in proportion to the light intensity falling on the surface of the photocell. In complete darkness the cell produces its greatest resistance.

There are many types of CdS cells on the market. One chooses a particular cell based on its dark resistance and light saturation resistance. The term *light saturation* refers to the state where increasing the light intensity to the CdS will not decrease its resistance any further. It is saturated. The CdS cell I used has approximately 100K-ohms resistance in complete darkness and 500 ohms of resistance when totally saturated with light. Under ambient light, resistance varies between 2.5K and 10K ohms.

This project requires two CdS cells. Test each cell separately. There may be an in-group variance that may change the scale factor used in each cell. In this project, I used a 0.022-μF capacitor, with the scale parameter set at 255 for both cells in the Pot command.

The schematic is shown in Fig. 6.30. The CdS cells are connected to port B, pins 2 and 3 (physical pin numbers 8 and 9). The photocells

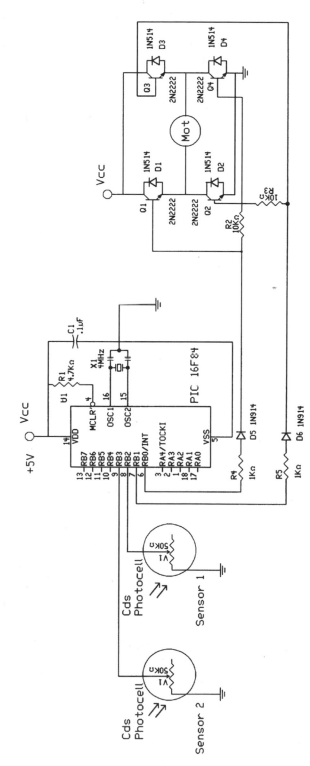

■ **6.30** *Schematic of light tracker circuit*

are mounted on a small piece of wood or plastic (see Fig. 6.31). Two small holes are drilled for each CdS cell for the wire leads to pass through. Longer wires are soldered onto these wires and connected to the PIC microcontroller.

One $\frac{3}{32}$″ to $\frac{1}{8}$″ hole is drilled for the gearbox motor's shaft. The sensor array is glued to the gearbox motor shaft (see Fig. 6.32).

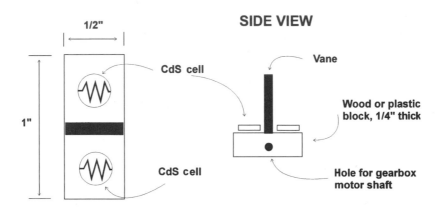

■ **6.31** *Construction on sensor array*

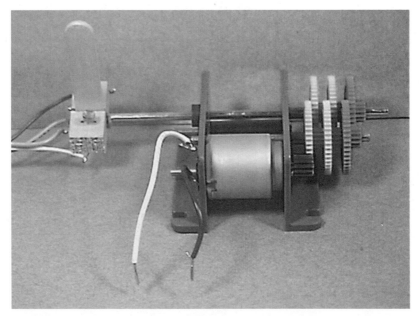

■ **6.32** *Photograph of sensor array on gearbox motor*

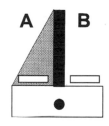 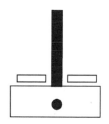 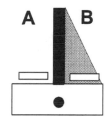

**A cell in shadow, tracker rotates to right**

**Equal illumination, no movement**

**B cell in shadow, tracker rotates to left**

## SIDE VIEW

■ **6.33** *Function of sensor array and pinpoint light source*

The operation of the tracker is shown in Fig. 6.33. When both sensors are equally illuminated, their respective resistances are approximately the same. As long as each sensor is within ±10 points of the other, the PIC program sees them as equal and doesn't initiate movement. This provides a group range of 20 points. This group range is the fuzzy part in fuzzy logic.

When either sensor falls in shadow, its resistance increases beyond our range and the PIC microcontroller activates the motor to bring both sensors under even illumination.

### DC motor control

The sun tracker uses a gearbox motor to rotate the sensor array toward the light source (see Fig. 6.34). The gearbox motor shown has a 4000:1 ratio. The shaft spins approximately 1 revolution per minute (rpm). You need a suitable slow motor (gearbox) to turn the sensor array.

The sensor array is attached (glued) to the shaft of the gearbox motor. The gearbox motor can rotate the sensor array clockwise

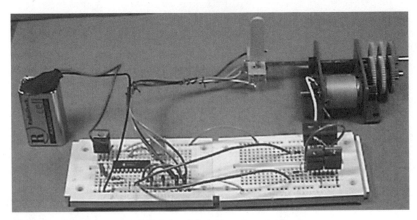

■ **6.34** *Photograph of finished light tracker projects*

(CW) and counterclockwise (CCW), depending upon the direction of current flowing through the motor.

To rotate the shaft (and sensor array) CW and CCW, we need a way to reverse current going to the motor. We will use what is known as an H-bridge. An H-bridge uses four transistors (see Fig. 6.35). Consider each transistor as a simple on/off switch as shown in the top portion of the drawing. It's called an H-bridge because the transistors (switches) are arranged in an H-type pattern.

When switches SW1 and SW4 are closed, the motor rotates in one direction. When switches SW2 and SW3 are closed, the motor rotates in the opposite direction. When the switches are opened, the motor is stopped.

The PIC microcontroller controls the H-bridge made of four TIP 120 Darlington NPN transistors; four 1N514 diodes; and two 10K-ohm, $\frac{1}{4}$-watt (W) resistors. Pin 0 is connected to transistors Q1 and Q4. Pin 1 is connected to transistors Q3 and Q4. Using either pin 0 or 1, the proper transistors are turned on and off to achieve CW or CCW rotation. The microcontroller can stop, rotate CW, or rotate CCW, depending upon the reading from the sensor array. Make sure the 10K-ohm resistors are placed properly or the H-bridge will not function.

The TIP 120 Darlington transistors are drawn in the schematic as standard NPN transistors. Many H-bridge circuit designs use PNP transistors on the high side of the H-bridge. The on resistance of PNP transistors is higher than that of NPN transistors. So in using NPN transistors exclusively in our H-bridge, we achieve a little higher efficiency.

## Diodes

Because the PIC is sensitive to electrical spikes (may cause a reset or lock-up), we place diodes across the collector-emitter junction of each transistor (Q1 to Q4). These diodes snub any electrical spikes caused by switching the motor windings on and off.

The PICBASIC compiler program is as follows:

```
'Fuzzy Logic Light Tracker Program
start:
low 0     'Pin 0 low
low 1     'Pin 1 low
pot 2,255,b0    'Read first CdS sensor
pot 3,255,b1     'Read second CdS sensor
if b0 = b1 then start    'If equal, do nothing
if b0 > b1 then greater    'If greater, check how much greater
if b0 < b1 then lesser    'If lesser, check how much lesser
greater:    'Greater routine
b2 = b0 - b1    'Find the difference
if b2 > 10 then cw    'Is it within range? If not, go to CW
goto start    'In range, do again
lesser:    'Lesser routine
b2 = b1 - b0    'Find the difference
if b2 > 10 then ccw    'Is it within range? If not, go to CCW
goto start    'Do again
cw:    'Turn the sensor array CW
high 0    'Turn on H-bridge
pause 100    'Let it turn for a moment
goto start    'Check again
ccw:    'Turn the sensor array CCW
high 1    'Turn on H-bridge
pause 100    'Let it turn a moment
goto start    'Check again
```

## Operation

When running, the light tracker will follow a light source. If both CdS cells are approximately evenly illuminated, the tracker does nothing. To test the light tracker, cover one CdS sensor with your finger. This should activate the gearbox motor and the shaft should begin to rotate.

If the shaft rotates in the opposite direction of the light source, reverse either the sensor input pins or the output pins to the H-bridge, *but not both.*

135

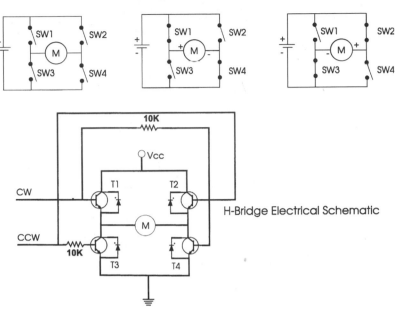

## H-Bridge Function

H-Bridge Electrical Schematic

■ **6.35** *H-bridge function and circuit schematic*

### Not fuzzy output

The output of our fuzzy light tracker is binary. The motor is either on or off, rotating CW or CCW. In many cases you would want the output to be fuzzy also. For instance, let us say you are making a fuzzy controller for an elevator. You would want the elevator to start and stop gradually (fuzzy) not abruptly as in binary (on/off).

Could we change the output of our light tracker and make it fuzzy? Yes. Instead of simply switching the motor on, we could feed a pulse-width modulation (PWM) signal that can vary the motor's speed.

Ideally the motor's speed would be in proportion to the difference (in resistance) of the two CdS cells. A large difference would produce a faster speed than would a small difference. The motor speed would change dynamically (in real time) as the tracker brings both CdS cells to equal illumination.

This output program may be illustrated using fuzzy logic graphics, groups, and membership sets. In this particular application creating a fuzzy output for this demonstration light tracker unit is overkill.

If you want to experiment, begin by using the `pulsout` and `pwm` commands to vary the DC motor speed.

### Neural sensors (logic)

With a small amount of programming, we can change our fuzzy logic sensors (CdS photocells) to neural sensors. Neural networks are an expansive topic. We will limit ourselves to one small example. For those who want to further pursue study into neural networks, I recommend a book I've written titled *Understanding Neural Networks* (Prompt, Indianapolis, 1998, ISBN 0-7906-1115-5).

To create neural sensors, we will take the numeric resistive reading from each sensor, multiply it by a weight factor, and then sum the results. The results are then compared to a tri-level threshold value (see Fig. 6.36).

Thus our small program and sensors are performing all the functions expected in a neural network. We may even be pioneering a neural first, by applying a multivalue threshold scheme. Do multivalue thresholds exist in nature (biological systems)? The answer is yes. For instance, an itch is an extremely low level of pain, and the sensation of burning can be felt when sensing something ice cold or hot.

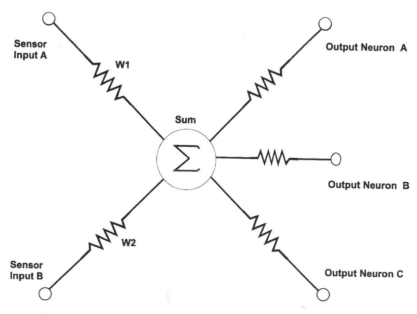

■ **6.36** *Schematic of simple tri-level neuron*

**Multivalue threshold**   Typically, in neural networks individual neurons have a singular threshold (positive or negative). Once the threshold is exceeded, the output of the neuron is activated. In our example the output is compared to multivalues, with the output going to the best fit.

Instead of thinking of the output as numeric values, think of each numeric range as a shape instead; a circle, square, and triangle will suffice. When the neuron is summed, it outputs a shape block (instead of a number). The receptor neurons (LEDs) have a shaped receiver unit that can fit in a shape block. When a shape block matches the receiver unit, the neuron becomes active (LED turns on).

In our case each output neuron relates to a particular behavior, sleeping, hunting, and feeding, all essential behaviors for survival in a photovore-style robot. Each output shape represents the current light level. At low light levels, the photovore stops hunting and searching for food (light). It enters a sleep or hibernation mode. At medium light level, the photovore hunts and searches for the brightest light areas. At high light levels, the photovore stops and feeds via solar cells to recharge its batteries.

Instead of building a photovore robot in this chapter, we will use an LED to distinguish between each behavior state (see Fig. 6.37). You can label the three LEDs sleeping, hunting, and feeding. Each LED will become active depending upon the light level received by the CdS cells.

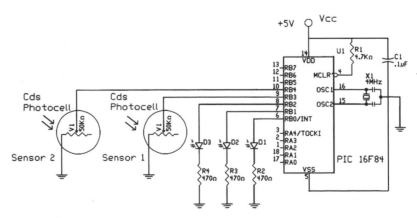

■ **6.37** *Schematic of basic neural circuit*

The program for PICBASIC compiler is as follows:

```
'Neural Demo
'Set up
low 0    'LED 0 off "Sleep"
low 1    'LED 1 off "Hunt"
low 2    'LED 2 off "Feed"
Start:
pot 3,255,b0    'Read first sensor
pot 4,255,b1    'Read second sensor
w2 = b0 * 3    'Apply weight
w3 = b1 * 2    'Apply weight
w4 = w2 + w3    'Sum results
'Apply thresholds
if w4 < 40 then feed    'Lots of light, feed
if w4 <= 300 then hunt    'Medium light, hunt
if w4 > 300 then snooze    'Little light, sleep
'Actions
feed:    'Feeding
low 0
low 1
high 2
goto start
hunt:    'Hunting
low 0
high 1
low 2
goto start
snooze:    'Sleeping * DON'T USE KEYWORD SLEEP *
high 0
low 1
low 2
goto start
```

## Parts list for programming the microcontroller

- ☐ PICBASIC compiler—$99.95
- ☐ PICBASIC Pro compiler (includes CodeDesigner Lite)—$249.95
- ☐ EPIC programmer—$59.95
- ☐ PICBASIC Pro compiler and EPIC programmer—$299.95
- ☐ PICBASIC compiler and EPIC programmer—$149.95
- ☐ CodeDesigner—hobbyist version—$45.00

- [ ] CodeDesigner—standard—$75.00
- [ ] (1) 16F84-4—$7.95*
- [ ] (1) 4.0-MHz crystal—$1.50*
- [ ] (2) 22-pF caps—$0.10 each*
- [ ] (1) 0.1-$\mu$F cap—$0.35*
- [ ] (1) 100-$\mu$F, 12V cap—$0.50*
- [ ] (1) 4.7K-ohm, $1/_4$-W resistor—$0.05*
- [ ] (2) 470-ohm, $1/_4$-W resistors—$0.05 each*
- [ ] (1) 7805 voltage regulator—$0.90*
- [ ] (2) Miniature LEDs—$0.25 each*
- [ ] (1) Solderless breadboard—$8.95*
- [ ] PIC-LED-02 PIC EXPERIMENTORS KIT [Consists of: (1) PIC16F84, (1) 4-MHz crystal, (2) 22-pF caps, (1) 10K-ohm, $1/_4$-W resistor, (1) 7805 voltage regulator, (1) solderless breadboard (2.1″ × 3.6″ 270 tie points), (8) 470-ohm resistors, (8) subminiature LEDs, (1) push-button switch, booklet with tutorial on binary number system, logic, and I/O of ports A and B]—$25.50
- [ ] 42-oz servo motor—$16.75

**140**

## Parts list for fuzzy light tracker and neural demonstration

- [ ] (2) CdS photocells
- [ ] (1) Flex sensor (nominal resistance 10K ohms)
- [ ] (2) 0.022-$\mu$F capacitors
- [ ] (1) 0.01-$\mu$F capacitor
- [ ] (4) TIP 120 NPN Darlington transistors
- [ ] (2) 10K-ohm resistors
- [ ] (6) 1N514 diodes
- [ ] (2) 1K-ohm resistor
- [ ] (1) 4000:1 gearbox motor

Parts available from:

Images Company

James Electronics

JDR MicroDevices

*These components plus additional components may be ordered as a single kit; see PIC-LED-02.

Radio Shack

Images SI, Inc.
39 Seneca Loop
Staten Island, NY 10314
(718) 698-8305
(718) 982-6145 (fax)

- ☐ (1) Solderless breadboard       Radio Shack PN# 276-175
- ☐ (1) 0.1-μF capacitor       Radio Shack PN# 272-1069
- ☐ (8) Red LEDs       Radio Shack PN# 276-208
- ☐ (8) 470-ohm resistors*       Radio Shack PN# 270-1115
- ☐ (1) 4.7K-ohm resistor       Radio Shack PN# 271-1124
- ☐ (8) 10K-ohm resistors       Radio Shack PN# 271-1126
- ☐ (1) 7805 voltage regulator       Radio Shack PN# 276-1770
- ☐ (2) 4-position PC-mounted switches   Radio Shack PN# 275-1301
- ☐ (1) 9V battery clip       Radio Shack PN# 270 325

Parts available from:

Radio Shack

James Electronics

JDR MicroDevices

*470-ohm resistors are also available in 16-pin dip package.

# 7

# Speech-controlled mobile robot

SPEECH IS AN IDEAL METHOD FOR ROBOTIC CONTROL AND communication. The speech-recognition circuit we will outline in this chapter functions independently from the robot's main intelligence [central processing unit (CPU)]. This is a good thing because it doesn't take any of the robot's main CPU processing power for word recognition. The CPU must merely poll the speech circuit's recognition lines occasionally to check if a command has been issued to the robot. We can even improve upon this by connecting the recognition line to one of the robot's CPU interrupt lines. By doing this, a recognized word would cause an interrupt, letting the CPU know a recognized word had been spoken. The advantage of using an interrupt is that polling the circuit's recognition line occasionally would no longer be necessary, further reducing any CPU overhead.

Another advantage to this stand-alone speech-recognition circuit (SRC) is its programmability. You can program and train the SRC to recognize the unique words you want recognized. The SRC can be easily interfaced to the robot's CPU.

Most voice recognition systems on the market today are software programs that require a host computer [usually an IBM personal computer (PC) or compatible] and a sound card. The speech-recognition system is still software-based even though it requires hardware (a sound card). These programs typically run in the background of a DOS or Windows environment, stealing themselves a portion of memory and CPU processing power while allowing other programs like Lotus or Word to run concurrently. The concurrent operation of the speech-recognition program slows the operation of any other program that runs using voice recognition.

There are many applications to voice recognition aside from robotics. Speech recognition will become the method of choice for controlling robots, virtual reality (VR), appliances, toys, tools, and computers. Because of the far-reaching potential of this technology, companies are developing speech recognition. The ability to control and command a computer (or appliance) by speaking directly to it will make working with that device easier, and more efficient and effective. At its most basic level, a speech-controlled device allows the user to perform parallel tasks, (i.e., hands and eyes can be busy elsewhere, while continuing to work with the computer or appliance).

There are three construction projects outlined in this chapter. The first project is a speech-recognition circuit. The second project interfaces the speech-recognition circuit to a mobile platform [radio-controlled (R/C) car]. The third project is a general interfacing board for the speech-recognition kit.

## Project 1: Programmable speech-recognition circuit

The first project is a programmable speech-recognition circuit. It is "programmable" in the sense that you can program the circuit to recognize up to 40 unique words of your own choosing. The heart of the circuit is a single integrated circuit (IC), the HM2007 speech-recognition chip. The chip provides the options of recognizing either 0.96-second (s) or 1.92-s word lengths.

Using 0.96-s word lengths enables the chip to recognize 40 independent words using an 8K × 8 static random access memory (RAM). You have the option to switch to the longer 1.92-s word length. While this reduces the word recognition count to 20 words, the longer word length can be used for phrases instead of isolated words. The circuit we will build here will use the 0.96-s word length, 40-word recognition library.

### Learning to listen

We take our speech-recognition abilities for granted. Listening to one person speak among several at a party is beyond the capabilities of today's speech-recognition systems. Speech-recognition systems like ours have a hard time separating and filtering out extraneous noise.

The operating distance one can speak from the microphone on the SRC without shouting is about 1 foot (ft). Because of this, when using the SRC on a mobile robot platform, we incorporate two small walkie-talkies. The output of one walkie-talkie is connected to the speech input of the SRC. The other walkie-talkie is used to speak to the robot via the SRC. This setup eliminates distance problems and extraneous noise.

Speech recognition is not speech understanding. Just because a computer can respond to a vocal command does not mean it understands the command spoken. Future voice-recognition systems will have the ability to distinguish nuances and meaning of words, to "Do what I mean, not what I say!" However, those systems are still years away.

## Speaker-dependent and speaker-independent speech recognition

Speech recognition is classified into two categories, speaker dependent and speaker independent. Speaker-dependent systems are trained by the individual who will be using the system. These systems are capable of achieving a high command count and better than 95 percent accuracy for word recognition. The drawback to this approach is that the system responds accurately only to the individual who trained the system. This is the most common approach employed in software for personal computers.

A speaker-independent system is trained to respond to a word regardless of who speaks. Therefore the system must be able to respond to a large variety of speech patterns, inflections, and enunciations of the target word. The number of command words is usually less than for the speaker-dependent system; however, high accuracy can still be maintained within processing limits. Industrial requirements more often need speaker-independent voice systems.

Our SRC will be speaker dependent. We can build in a little speaker independency by allocating more than one word space to a target word and then programming different word enunciations in the allocated spaces. Each of these word spaces would trigger the same command.

## Recognition style

Speech-recognition systems have another constraint concerning the style of speech they can recognize. There are assumed to be three styles of speech: isolated, connected, and continuous.

### Isolated

These speech-recognition systems can just handle words that are spoken separately. This is the most common speech-recognition system available today. The user must pause between each word or command spoken. Our speech-recognition circuit will use isolated words.

### Connected

This is a halfway point between isolated word and continuous speech recognition. It allows users to speak multiple words. The HM2007 can be set up to identify words or phrases 1.92 s in length. This reduces the word recognition dictionary number to 20.

### Continuous

This is the natural conversational speech we use in everyday life. It is extremely difficult for a recognizer to shift through the text since the words tend to merge together. For instance, "Hi, how are you doing?" sounds like "Hi, howyadoin." Continuous speech-recognition systems are on the market and are under continual development.

## Building the speech-recognition circuit

The demonstration circuit operates in the HM2007's manual mode. This mode uses a simple keypad and microphone to program the HM2007 chip.

### Keyboard

The keyboard is a telephone keypad made up of 12 normally open switches.

| 1 | 2 | 3 |
|-------|---|-------|
| 4 | 5 | 6 |
| 7 | 8 | 9 |
| Clear | 0 | Train |

When the circuit is turned on, the HM2007 checks the onboard static RAM. If the RAM checks out, the board displays "00" on the seven-segment display chips, lights the red light-emitting diode (LED) (Ready), and waits for a command.

### To train

Press "1" (display will show "01") and the LED will turn off. Then press "T" (Training), and the LED will turn back on.

Hold the microphone close to your mouth and say the training word. For instance, let's use the word "computer" as the training word. Say the word computer into the microphone. If the circuit accepts the word spoken, the LED will blink. The word computer is now programmed as the "01" word. Whenever the circuit hears the word computer, it will display "01" on its digital output.

If the LED did not blink when you said the word computer, either repeat the word again louder or start over completely by pressing "01" then "T."

Continue training new words in the circuit. Press "02" then "T" to train the second word. The circuit will accept up to 40 words. You do not need to train all 40 words to use the circuit. Train the circuit with just the words you need and start using the circuit.

### Testing recognition

Repeat a trained word into the microphone. The number of the word should be displayed on the segmented display. For instance, if the word "directory" was trained as word number 25, saying the word "directory" into the microphone will cause the number 25 to be displayed.

### Error codes

The chip provides the following error codes:

- ☐ 55 = word too long
- ☐ 66 = word too short
- ☐ 77 = no word match found

### Clearing memory

You can erase individual words in memory by entering the word number you want to erase and hitting the CLR button. To erase all the words in memory, press "99" and then CLR.

### More about the HM2007 chip

The HM2007 is a single-chip complementary metal-oxide semiconductor (CMOS) voice-recognition large-scale integration (LSI) circuit. The chip contains an analog front end, voice analysis, recognition, and system control functions. The chip may be used in a stand-alone or connected CPU.

### Features

- ☐ Single-chip voice-recognition CMOS LSI
- ☐ Speaker-dependent
- ☐ External RAM support
- ☐ Maximum of 40-word recognition
- ☐ Maximum word length of 1.92 s

☐ Microphone support

☐ Manual and CPU modes available

☐ Response time less than 300 milliseconds (ms)

☐ 5 volt (5V) power supply

### Circuit construction

The speech-recognition circuit is available in kit form from Images Company (see the parts list at the end of the chapter). The schematic is shown in Fig. 7.1. The components can be mounted and wired on a standard printed circuit board (PCB).

Solder the keypad to the board according to Fig. 7.2. You will have just seven wires from the keypad to the HM2007 on the PCB. The number next to each wire coming out of the keypad refers to the pin number it's connected to on the HM2007 IC.

Figure 7.3 shows the top view of the parts placement on the PCB. Figure 7.4 is the complete speech-recognition circuit.

### Independent recognition system

This demonstration circuit allows you to experiment with dependent as well as independent systems. The system is typically trained as speaker dependent, meaning the voice that trained the circuit also uses it.

We will take the other track and train the system for speaker independent recognition. To accomplish this we will use four word spaces for each target word.

To simplify the digital logic, the allocation of word spaces is as follows. Our circuit will only look at the first [least significant digit (LSD) on the display] digit space for recognition. This means that the word spaces 01, 11, 21, and 31 will all be recognized as the same word. Since we are only decoding the first digit, they all look like word space 1. Likewise word spaces 04, 14, 24, and 34 all look like word space 4.

This system works most of the time, but a problem is encountered when an error code pops up.

☐ 55 = word too long

☐ 66 = word too short

☐ 77 = no word match found

Obviously the base circuit would identify these error codes as word 5, 6, and 7, respectively. There are two ways to work around this problem. The first way is to use a dedicated logic circuit

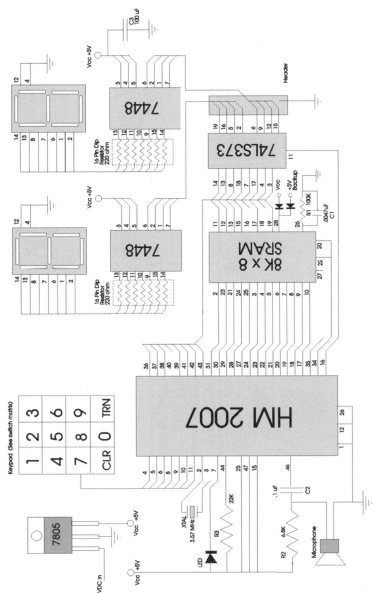

■ 7.1 Schematic of speech-recognition circuit

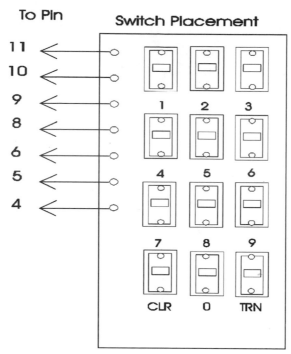

**7.2** *Keypad wiring to speech-recognition circuit*

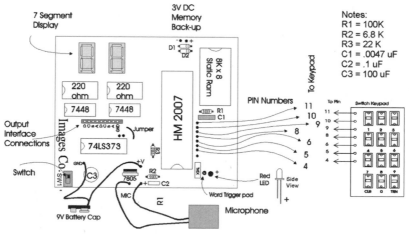

**7.3** *Top view of parts placement on PCB*

(see Fig. 7.5) that brings a line high when the digits 5, 6, or 7 appear in the most significant byte (MSB). This line becomes an enable-disable line. This circuit brings the line high when the digits 5, 6, or 7 are displayed, so an interface should interpret this line going high as a disable.

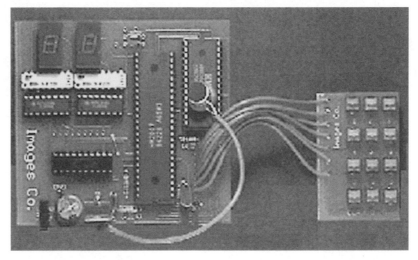

■ **7.4** *Finished speech-recognition circuit*

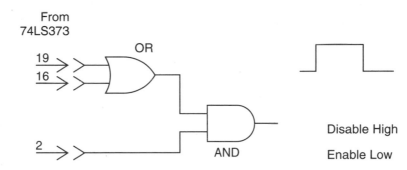

From
74LS373

OR

19

16

2

AND

Disable High

Enable Low

Error Code Detection from MSB of 74LS373

■ **7.5** *Error detection circuit taken from upper BCD number*

The second way to work around the problem is to simply use a PIC microcontroller to read the entire 8-bit output from the SRC. Any word number above 40 is an error and should be ignored. While we are not interfacing this circuit to a microcontroller here, it should be evident to anyone who has worked with the PICBASIC compiler and PIC chips (see Chap. 6) in other applications that this interface would not present a problem. In Chap. 15 we use a PIC inside the speech-controlled circuit for controlling a robotic arm.

The 8-bit output is taken from the output of the 74LS373 data octal latch. The output is not a standard 8-bit byte, but it is broken into two 4-bit binary coded decimal (BCD) nibbles. BCD code is related to standard binary numbers as Table 7.1 illustrates.

**■ Table 7.1**

| Decimal | Binary | BCD |
|---------|--------|-----|
| 0 | 0000 | 0000 |
| 1 | 0001 | 0001 |
| 2 | 0010 | 0010 |
| 3 | 0011 | 0011 |
| 4 | 0100 | 0100 |
| 5 | 0101 | 0101 |
| 6 | 0110 | 0110 |
| 7 | 0111 | 0111 |
| 8 | 1000 | 1000 |
| 9 | 1001 | 1001 |
| 10 | 1010 | 0001 0000 |
| 11 | 1011 | 0001 0001 |
| 12 | 1100 | 0001 0010 |
| 13 | 1101 | 0001 0011 |
| 14 | 1110 | 0001 0100 |
| 15 | 1111 | 0001 0101 |
| 16 | 0001 0000 | 0001 0110 |
| 17 | 0001 0001 | 0001 0111 |
| 18 | 0001 0010 | 0001 1000 |
| 19 | 0001 0011 | 0001 1001 |
| 20 | 0001 0100 | 0010 0000 |

As you can see, the binary and BCD numbers remain the same until reaching decimal 10. At decimal 10, BCD jumps to the upper nibble and the lower nibble resets to zero. The binary numbers continue to decimal 15, and then jump to the upper nibble at 16 where the lower nibble resets. If a computer is expecting to read an 8-bit binary number and BCD is provided, this will be the cause of errors.

## Project 2: Interface circuit

The interface circuit revolves around the 4028 BCD integrated circuit. The 4028 takes the lower BCD output from the 74LS373 on the speech-recognition board and outputs a high signal; see the truth table on 4028, Table 7.2.

The schematic for the interface circuit is shown in Fig. 7.6. The inputs A, B, C, and D to the 4028 are the lower BCD numbers from the 74LS373. When I stripped the car of its radio-control (R/C) equipment, I was left with a group of wires that, when powered, performed the basic driving functions. The robot car has just four functions: forward straight, forward right, forward left, and reverse.

| Input | | | | Output | | | | | | | | | |
|---|---|---|---|---|---|---|---|---|---|---|---|---|---|
| D | C | B | A | Q9 | Q8 | Q7 | Q6 | Q5 | Q4 | Q3 | Q2 | Q1 | Q0 |
| 0 | 0 | 0 | 0 | 0 | 0 | 0 | 0 | 0 | 0 | 0 | 0 | 0 | 1 |
| 0 | 0 | 0 | 1 | 0 | 0 | 0 | 0 | 0 | 0 | 0 | 0 | 1 | 0 |
| 0 | 0 | 1 | 0 | 0 | 0 | 0 | 0 | 0 | 0 | 0 | 1 | 0 | 0 |
| 0 | 0 | 1 | 1 | 0 | 0 | 0 | 0 | 0 | 0 | 1 | 0 | 0 | 0 |
| 0 | 1 | 0 | 0 | 0 | 0 | 0 | 0 | 0 | 1 | 0 | 0 | 0 | 0 |
| 0 | 1 | 0 | 1 | 0 | 0 | 0 | 0 | 1 | 0 | 0 | 0 | 0 | 0 |
| 0 | 1 | 1 | 0 | 0 | 0 | 0 | 1 | 0 | 0 | 0 | 0 | 0 | 0 |
| 0 | 1 | 1 | 1 | 0 | 0 | 1 | 0 | 0 | 0 | 0 | 0 | 0 | 0 |
| 1 | 0 | 0 | 0 | 0 | 1 | 0 | 0 | 0 | 0 | 0 | 0 | 0 | 0 |
| 1 | 0 | 0 | 1 | 1 | 0 | 0 | 0 | 0 | 0 | 0 | 0 | 0 | 0 |

Each function is controlled by an electric motor or motor/solenoid combination that can be powered by an NPN transistor. So each function requires one NPN transistor. Four transistors connected to the outputs of the 4028, labeled Q1 to Q4, control this electric car.

For purposes of a clear illustration, Fig. 7.6 shows only one NPN transistor connected to output Q1 powering an electric motor. The exact model R/C car I used when writing this chapter is no longer available. Even so, most inexpensive R/C cars will function in a similar manner. Remove the R/C equipment from the car. You will be left with wires leading to the drive motor that will either have to be grounded or connected to Vcc to power the motor. Turning left and right is usually accomplished with an inexpensive solenoid. Again check the wires from the steering solenoids to see if they need to be connected to ground or connected to Vcc.

## Walkie-talkies

Radio Shack sells a number of inexpensive walkie-talkies. Since the operating distance of the microphone on the SRC is about 1 ft, using a pair of walkie-talkies increases the distance with which one can operate the mobile robotic platform via the SRC. The speaker output of the walkie-talkie is connected to pin 46 on the HM2007 through capacitor C1. Capacitor C1 will block any DC component output from the walkie-talkie.

## Acoustic coupling

If you don't want to take apart and solder wires between the walkie-talkie and SRC directly, you can try an acoustic coupling.

153

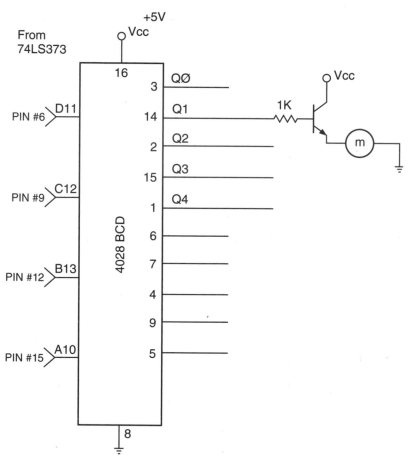

**■ 7.6** *Interface circuit to modified R/C car*

Essentially you tape the microphone of the speech-recognition kit to the speaker of the walkie-talkie. The microphone and speaker assembly may be enclosed in a box by itself to reduce background noise.

### Training and controlling the mobile robot

The SRC on the mobile robotic platform should be trained using the walkie-talkie, if that's the way it will be operated and used. The digital display on the board is active when the interface board is connected, so it can be used to check word recognition accuracy. Find out the range of the walkie-talkie system. Don't let the mobile robot travel outside of this range or you will end up running after it yelling "stop, stop, stop" into the walkie-talkie. Controlling the robot is as simple as talking to it, and it's pretty impressive to boot.

■ **7.7** *Speech-controlled R/C car*

### New board features

The voice-controlled mobile robotic platform is shown in Fig. 7.7. The circuit board on the robot looks a little different than that in Fig. 7.4. The reason for this is that I happened to get a prototype of the latest revision of the speech kit.

The latest revision makes interfacing to the SRC easy. There are nine PC holes [for a pin header (two 4-bit nibbles plus ground)] that connect to the output of the onboard 74LS373. The output of the onboard 74LS373 is the upper BCD used for word-error detection and the lower BCD used to activate the 4028. A trigger signal is available by the red LED.

In addition to the interface hookups, the board has a 3V input for memory backup. This makes the static RAM on the speech board nonvolatile. So you can turn the board on and off without losing the words programmed in the static RAM. In the original version, when you turned off the power, you lost the words programmed in the RAM.

## Project 3: General speech-recognition interfacing circuit

The speech interface to the mobile R/C car is a specialized application. The next interface circuit (see Fig. 7.8) is a more general circuit and lends itself to controlling a variety of devices that include robots, electric circuits, and appliances.

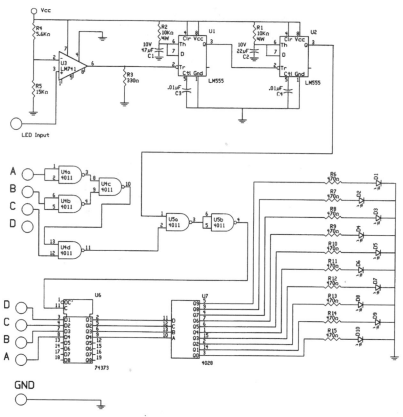

■ **7.8** *General interface circuit for speech-recognition circuit*

To keep the interface circuit from becoming too large and at the same time to enhance the robustness and accuracy of the SRC, we will again limit the interface to control 10 on/off switches. If you need the full 40-word vocabulary, you can design the interface circuit by expanding the circuit ideas illustrated in this chapter. Using just 10 on/off switches allows us to use four word spaces for each target (command) word, as before. Each of the four word spaces assigned to a target word will hold a slightly different enunciation of the target word. With four different enunciations of each target word, the SRC becomes more robust and word recognition accuracy increases.

We choose the word spaces as before, so that the LSD of any four target (command) word spaces is the same. An example will make this programming scheme clear.

Suppose we are making a voice control for an electric wheelchair. We decide to use the following list of command (target) words:

- ☐ Forward
- ☐ Backward
- ☐ Left
- ☐ Right
- ☐ Stop
- ☐ Sleep
- ☐ On
- ☐ Lock
- ☐ Unlock
- ☐ Stop (The command stop is so important in this application that it may take up more that one command position.)

The first command we want to train the circuit to recognize is forward. We will use the following four word spaces: 10, 20, 30, and 40. By dropping the most significant digit (MSD) of each number, we are left with the LSD that is the same for all four word spaces, word number 0. Similarly the next command word, backward, will use word spaces 01, 11, 21, and 31. Dropping the MSD again, we are left with word number 1.

The interface must recognize the word error codes and not mistake them for word numbers 5, 6, and 7. The circuit uses two 4011 NAND ICs configured to operate as OR and AND gates (as shown in Fig. 7.8) to detect the 55, 66, and 77 word errors.

## Connection to speech kit

The speech-recognition kit has nine solder holes between the 74LS373 and 7448 chips for connecting an interface circuit (see Fig. 7.8). Eight lines represent the two BCD numbers, and the ninth pad is a ground. There is also one open pad by the red LED. A wire soldered here is used as an input signal to a word trigger for the interface circuit.

## How it works

To begin, the interface circuit must be able to react whenever the SRC hears a word. When the SRC hears a word, it attempts recognition and the red LED blinks off momentarily.

The current to the LED is used as a word trigger. To use this as a trigger, we set up a comparator connected to the cathode side of the LED. The reference voltage for the comparator is set at 3.64 V using a voltage divider consisting of two resistors, 5.6K- and 15K-ohms.

The output of the comparator is normally high. When triggered by the 4.5V pulse from the LED line, it outputs a negative pulse trigger. For a comparator we are using a standard 741 operational amplifier (op-amp).

Speech recognition can take up to 300 ms. During this time delay, the BCD outputs remain stable and do not change. If our interface operates too quickly, it will already be finished updating the output before the SRC has a chance to update the BCD output.

To prevent this from occurring, we delay the negative pulse trigger by sending it through two 555 timers (or one 556 timer) set up in mono-stable mode. The negative pulse from the comparator initiates a 470-ms output pulse from the first timer, which is connected to the second timer. The second timer outputs a 220-ms pulse.

The 470-ms pulse allows more than enough time for the new BCD numbers to be outputted. When the 470-ms pulse from the first timer goes low, it initiates an output pulse from the second timer. This output is a positive pulse lasting approximately 220 ms. During this time, the interface output is updated provided the error code detector (ECD) is outputting a logic high.

The second timer (220-ms pulse) output is connected to one input of an AND gate. The other leg of the AND gate is connected to two other gates (NAND and OR) that make up our ECD. The ECD is connected to the most-significant-digit BCD number. Whenever the BCD number is equal to 5, 6, or 7, the ECD outputs a logic low. For all other numbers it outputs a logic high. When the output of the ECD is positive, the positive pulse from the second timer allows the least-significant-digit BCD number to propagate through to the output of the interface circuit.

The output high from the ECD combined with the positive pulse from the timer triggers a logic high from the AND gate that enables the 74LS373 (data octal latch). With the 74LS373 enabled, any number outputted on the lower BCD number propagates through the 74LS373 and is latched. The four outputs of the 74LS373 are connected to the inputs of a 4028 BCD-to-decimal decoder.

On the other hand, when the output of the ECD is low, which happens when the numbers 5, 6, and 7 are outputted, the corresponding input to the AND gate is low. With this AND input kept low, when the positive pulse from the second timer arrives, the output of the AND gate will remain low, thereby keeping the 74LS373 disabled and not allowing the lower BCD number to propagate through to the 4028.

This is how the error detector prevents numbers 55, 66, and 77 from being mistaken for words 5, 6, and 7.

When the BCD propagates through the 74LS373, it connects to the input of a 4028 BCD-to-decimal decoder. The 4028 reads the BCD number and outputs a logic high (+5V) on the appropriate decimal output line (0 to 9).

## Creating a more useful output

The logic high output from the 4028 can be used to control alternating current (AC) and direct current (DC) loads. However, it is much better to run the output from the 4028 through a flip-flop first. The reason is that the 4028 by itself will only keep one of its outputs high at any given moment. So whenever the circuit turns something on, whatever may have been turned on will be turned off. Not very convenient. The flip-flop solves this problem. Once triggered by a logic high, the flip-flop output will remain high until a second signal (low-high) brings its output low. The result of using a flip-flop on an output line is twofold.

Primarily one can turn on and off any number of outputs without affecting the status of any other line. Secondly, the same command may be used to turn on (first time spoken) and then turn off (second time spoken) a circuit. So instead of having two separate on/off commands for each device (like light-on and light-off), the same command can be used a second time to turn off the device (light-light). In some cases this is like doubling your command vocabulary.

Figure 7.9 shows a 4013 flip-flop circuit that may be used. Each 4013 IC has two usable flip-flops. The inputs of the flip-flops are connected directly to the line output of the 4028.

Figure 7.10 shows one circuit and two fragments that may be connected to the flip-flop for controlling different types of loads. The circuit in *A* is an NPN Darlington transistor, with a DC source and resistive load. One may also use this type of circuit to open and close a relay, as shown in *B*. The relay can control AC and DC loads (resistive or inductive). In *C* the output of the 4013 is connected to an optocoupler with a Triac output.

## Operation

The speech-recognition circuit is trained as described previously. With the interface circuit connected, each word command will light an LED or circuit, depending upon what's connected to the output of the 4028 BCD-to-decimal decoder chip.

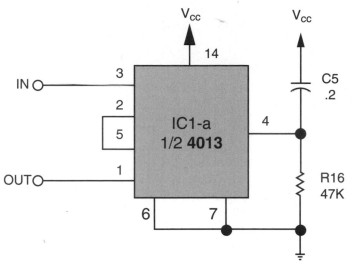

■ **7.9** One-half of 4013 flip-flop used for signal latching. Connected directly to an output on the 4028 in Fig. 7.8, and used to prevent an active circuit from turning off when another is turned on

## Improving recognition

There are a number of techniques one can employ to improve or optimize recognition. Word selection is the primary one. Avoid homonyms, words that sound alike, for instance, red, bed, said, and dead. To optimize recognition use dissimilar sounding words. In many cases a synonym or approximate synonym can be used in place of a word. For instance use "crimson" or "scarlet" in place of "red." For "bed" try using "bunk," "mattress," "berth," or "cot." For "said" one may use "spoke," "voiced," or "uttered." For "dead" one could use "deceased," "expired," or "late." A little thought will solve any homonym dilemma.

## Match environment and equipment

1. *Distance.* The distance the microphone is away from the speaker's mouth should be approximately the same for training and recognition.

2. *Stress.* The voice changes under stress or excitement. For instance, if you are creating a voice-controlled joystick to fly your favorite military flight simulator, your voice when engaged in a dogfight yelling "Fire! Fire! Bank Left" will be quite different than when sitting at your desk calmly programming your voice into the chip. Therefore, you must try to emulate the stress and excitement you will feel when playing the game when programming the voice commands.

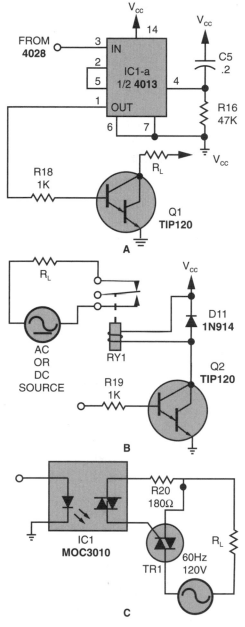

■ **7.10** *These circuits can be connected to the interface circuit to allow the voice-recognition system to control different types of loads. The circuit in A uses a flip-flop to latch a Darlington transistor in the on or off condition. By replacing the Darlington in A with the relay configuration in B, the circuit can be used to control high-current resistive or inductive AC or DC loads. The circuit in C, when combined with the flip-flop in A, provides isolation between the load and controlling circuit, while allowing you to latch the AC load on or off*

*Speech-controlled mobile robot*

3. *Exertion.* Physical stress is another factor. If you are programming exercise equipment (a Stairmaster or stationary bike) to respond to voice, you might want to record people who are a little out of breath.

4. *Background noise.* Background noise is always a problem. As stated previously, a steady background noise (air conditioner) has less impact on speech recognition accuracy than a nonsteady background noise (TV).

### Speech-controlled robotic arm

In Chap. 15, another derivative of the speech-recognition interface is used to control a robotic arm.

## Parts list for speech-recognition circuit

- ☐ (1) IC1 HM2007 IC
- ☐ (1) IC2 SRAM 8K X 8
- ☐ (1) IC3 74LS373
- ☐ (2) IC4 and IC5 7448
- ☐ (1) XTAL 3.57 MHz
- ☐ (1) Speech-recognition PCB
- ☐ (1) 12-contact keypad
- ☐ (2) 7-segment displays
- ☐ (2) 16-pin, 220-ohm, $\frac{1}{4}$-W resistor packs
- ☐ (1) 22K-ohm, $\frac{1}{4}$-W resistor
- ☐ (1) 5.6K-ohm, $\frac{1}{4}$-W resistor
- ☐ (1) 0.0047-μF cap
- ☐ (1) C2 100-μF, 16V cap
- ☐ (1) C5 0.1-μF cap
- ☐ (1) 7805 voltage regulator
- ☐ (1) Microphone
- ☐ (1) 9V battery clip
- ☐ Complete speech-recognition kit—$100.00

## Parts list for interface circuit

- ☐ (2) 4011 Quad 2 input NAND
- ☐ (1) 74LS373 Octal D flip-flop tri-state
- ☐ (1) 4028 BCD-to-decimal decoder

- ☐ (2) 555 timers
- ☐ (1) LM741 op-amp
- ☐ (1) 5.6K-ohm resistor
- ☐ (1) 15K-ohm resistor
- ☐ (1) 330-ohm resistor
- ☐ (2) 10K-ohm resistors
- ☐ (10) 470-ohm resistors
- ☐ (1) 47-μF cap
- ☐ (1) 22-μF cap
- ☐ (2) 0.01-μF caps
- ☐ (10) Miniature LEDs
- ☐ Optional: 4013 dual-type D flip-flops, TIP 120 NPN Darlington transistors

Parts available from:

Images Company
39 Seneca Loop
Staten Island, NY 10314

http://www.imagesco.com

# 8

# Behavioral-based robotics, neural networks, nervous nets, and subsumption architecture

ROBOTS OF THE CLASS DISCUSSED IN THIS CHAPTER DO NOT have a central processing unit (CPU). Rather they function on a neural stimulus-response mechanism.

The robotic stimulus-response mechanism goes by a number of names, including neural network, behavioral-based robotics, subsumption architecture, and nervous network. William Grey Walter pioneered behavioral-based robotics in the late 1940s. Independent of Walter's work, neural-based robotic response was academically explored and developed in the 1980s by Valentino Braitenberg in his book *Vehicles: Experiments in Synthetic Psychology*. Rodney Brooks at the Massachusetts Institute of Technology (MIT), inspired by work accomplished by Walter, developed his own derivative of stimulus responses he calls "subsumption architecture." Mark Tilden, inspired by work done by Rodney Brooks, founded BEAM robotics, which uses "nervous nets."

Behavioral-based robotics is a hot topic and one that will continue to get hotter in the future. In these architectural schemes the stimulus-response mechanisms can be layered on top of one another. A multilayer stimulus-response mechanism can exhibit what appears as strikingly intelligent behavior as in the intelligent photovore robot discussed later.

For the time being, I will use the name "behavioral-based" to describe this stimulus-response mechanism. The behavioral-based approach is one of the two main approaches to implementing intelligence in robots as discussed in Chap. 6. One approach is called "top-down intelligence" and the other is called "bottom-up intelligence."

To implement intelligent control functions in a mobile robot [by using the term "functions" I am limiting the field of discussion to the movement (mobility) and exploration of an environment for simplicity, but this is by no means a real restriction on either approach discussed], one must decide on which approach is better to accomplish the task. The top-down approach attempts to create an expert system or program to perform a controlled search and discover. The bottom-up approach creates "artificial" behavior in the robot and then causes it to explore and discover.

At first glance you may not see much of a difference in either approach, but there is one and it's quite significant. If the expert system approaches a situation (or terrain) it hasn't been programmed to handle, it will falter. The behavior system on the other hand isn't looking for any template "programmed" situation to calculate procedures and couldn't care less about the situation; it just goes on exploring.

Robotists have found over the last 30 years of experimentation that bottom-up programming (behavioral-based) is successful many times where top-down programming fails.

## Robotics pioneer

As previously stated, one of the first pioneers in the bottom-up approach to robotics was William Grey Walter. He was born in Kansas City, Missouri, in 1910. When he was 5 years old, his family moved to England. He attended school there, graduating from King's College, Cambridge, in 1931. After graduation, he began doing basic neurophysiological research in hospitals.

Early in his career, Walter found interest in the work of the Russian psychologist Ivan Pavlov, famous for his stimulus-response experiment with dogs. In the experiment, Pavlov rang a bell just before providing food for his dog subjects. After a while, the dogs became conditioned to salivate just by hearing the bell.

Another contemporary of Walter, Hans Berger, invented the electroencephalograph (EEG) machine. When Walter visited Berger's laboratory, he saw refinements he could make to Berger's EEG machine. When he did so, the sensitivity of the EEG machine was improved, and new EEG rhythms below 10 hertz (Hz) could now be observed in the human brain.

Walter's studies of the human brain led him to study the neural network structures in the brain. The vast complexities of the biological networks were too overwhelming to map accurately or replicate. Soon he began working with individual neurons and the electrical equivalent of a biological neuron. He wondered what type of behavior could be created using just a few neurons.

To answer this question, in 1948 Walter built a three-wheeled turtle-like mobile robot. The mobile robot measured 12″ high and about 18″ long. What is fascinating about this robot is that it used just two electronic neurons but exhibited interesting and complex behaviors. The first two robots were affectionately named Elmer and Elsie (ELectroMEchanical Robot, Light Sensitive). Walter later renamed the style of robots Machina Speculatrix after observing the complex behavior they exhibited.

Remember, in the 1940s the transistor had not yet been invented, so the electronic neurons for the robot were made using vacuum tubes. Vacuum tubes consume considerably more power than semiconductors, and so the original robot was fitted with a rather large rechargeable battery.

The robot's reflex or nervous system consisted of two sensors connected to two neurons. One sensor was a light-sensitive resistor and the other sensor was a bump switch connected to the robot's outer housing.

The three wheels of the robot were in a triangular configuration. The front wheel had a motorized steering assembly that could rotate a full 360 degrees in one direction. In addition, the front wheel also contained a drive motor for propulsion. Since the steering could continually rotate a full 360 degrees, the drive motor's electric power came through slip rings mounted on the wheel's shaft.

The photosensitive resistor was mounted onto the shaft of the front wheel steering-drive assembly. This ensured that the photosensitive resistor was always facing in the direction that the robot was moving.

## Four modes of operation

While primarily of a photovore (light-seeking) type, the robot exhibited four modes of operation. It should be mentioned that the robot's steering motor and drive motor were usually active.

☐ *Search.*   Ambient environment is at a low light level or darkness. Robot's response: steering motor on full speed, drive motor on one-half speed.

☐ *Move.*   Found light. Robot's response: steering motor off, drive motor full speed.

☐ *Dazzle.*   Bright light. Robot's response: steering one-half speed, drive motor reversed.

☐ *Touch.*   Hit obstacle. Robot's response: steering full speed, reverse drive motor.

## Observed behavior

In the 1950s, Walter wrote two *Scientific American* articles ("An Imitation of Life," May 1950; "A Machine That Learns," August 1951) and later a book titled *The Living Brain* (Norton, New York, 1963). The interaction between the neural system and the environment generated unexpected and complex behaviors.

In one experiment Walter built a hutch, where Elsie could enter and recharge its battery. The hutch was equipped with a small light that would draw the robot to it as the robot's batteries ran down. The robot would enter the hutch and its battery would automatically be recharged. Once the battery recharged, the robot would leave the hutch to search for new light sources.

In another experiment Walter fixed small lamps on each tortoise shell. The robots developed an interaction that to an observer appeared like a kind of social behavior. The robots danced around each other, at times attracted and then repelled, reminding one of a robotic mating ritual or territorial-marking behavior.

## Building a Walter tortoise

We can imitate most functions in Walter's famous tortoise. The program we will use simulates the neurons used in the original robot. To fabricate the chassis, we need to do a little metalwork. Working metal is made a lot easier with the following tools:

☐ *Center punch.*   Used to make a dimple in sheet metal to facilitate drilling. Without the dimple, the drill is more likely to "walk off" the drill mark. To use, hold the center punch in the center of the hole needing to be drilled. Hit the center punch with the hammer to make a dimple.

- [ ] *Hand shears.* Used to cut sheet metal. I would advise the purchase of 14″ metal shears. Use like scissors to cut metal. Note: Metal is a lot harder to cut than paper.
- [ ] *Nibbler.* Used to remove (nibble) small bits of metal from a sheet. Used to nibble cutouts and square holes in light-gauge sheet metal. Note: Radio Shack sells an inexpensive nibbler.
- [ ] *Vise.* Used to hold metal for drilling and bending.
- [ ] *Drill*
- [ ] *Hammer*

Most hardware stores will carry these simple metalworking tools. They will also carry the light-gauge sheet metal and aluminum bar needed to make the chassis.

I built my chassis out of $\frac{1}{8}″ \times \frac{1}{2}″$ aluminum rectangle bar and 22- to 24-gauge stainless-steel sheet metal. Stainless steel is harder to work with than cold-rolled steel (CRS), and if I had to do it over again, I would use aluminum or CRS.

### Drive and steering motors

The drive motor is a 100:1 gearbox motor (see Fig. 8.1). I like this gearbox motor because it has a motor mounting bracket. For the steering motor I used a standard 42-ounce (oz) torque servo motor. There are three pieces of sheet metal one needs to fabricate.

■ **8.1** *100:1, 1.5- to 3.0 - VDC gearbox motor*

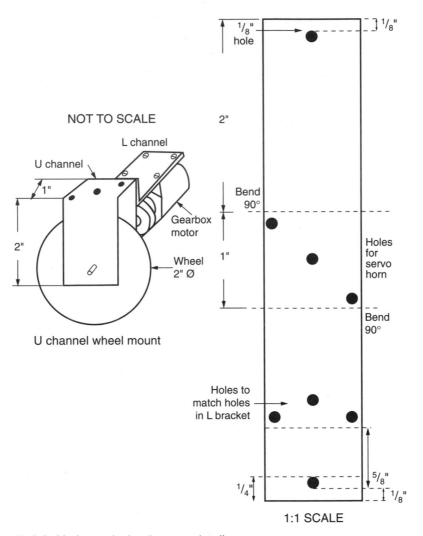

NOT TO SCALE

2"

1/8" hole

1/8"

L channel

U channel

1"

2"

Gearbox motor

Wheel 2" Ø

U channel wheel mount

Bend 90°

1"

Holes for servo horn

Bend 90°

Holes to match holes in L bracket

5/8"

1/4"

1/8"

1:1 SCALE

■ **8.2** *U channel wheel mount detail*

The U channel (see Fig. 8.2) holds the front wheel and drive motor. The U bracket is fabricated from 22-gauge, 1″ × 5″ sheet metal. Three holes in the center area are drilled to mount the servo horn from the servo motor. The center drill hole ($^1/_8$″) is larger than the two outer holes ($^1/_{16}$″). Remove the servo horn from the servo motor by unscrewing the center screw and pulling straight up on the horn. Line up the servo horn on the bracket and mark the center and two outer holes. Drill the three holes. Mount the servo horn, using the center servo motor screw. For the outer holes use two 0-80 machine screws and nuts. Drill three $^1/_8$″ holes for mounting the L bracket to the side. Drill the two axle holes for the front wheel. Use $^1/_8$″ holes.

Mount the metal in a vise and make the two 90 degree bends at the marked lines to form the metal into a U bracket.

Use the L bracket to mount the drive motor to the U channel (see Fig. 8.3). The L bracket is 1.5″ × 3″. Use the gearbox motor to mark the mounting holes for the gearbox. Make sure the three holes in the L bracket for mounting to the U bracket match the mating holes in the U bracket.

Figure 8.4 is a diagram of the base with a diagram for the 42-oz servo motor. The base measures 3″ × 5.5″. The base will hold the power supply and the electronics. Use the servo motor diagram to remove metal from the base.

First drill the four ($\frac{1}{8}$″) holes. Next use the drill to cut holes all along the rectangle inside the perimeter of the servo motor hole. Removing metal this way is much easier than trying to saw or nibble it away. When you have removed as much material as possible this way, use the metal nibbler to finish the job. Before mounting the servo, file the edges of the hole. Drill the two back holes for the rear axle bracket.

The rear axle bracket is shown in Fig. 8.5. It is made from $\frac{1}{8}$″ × $\frac{1}{2}$″ × 10″ aluminum bar. Drill the four $\frac{1}{8}$″ holes in the aluminum be-

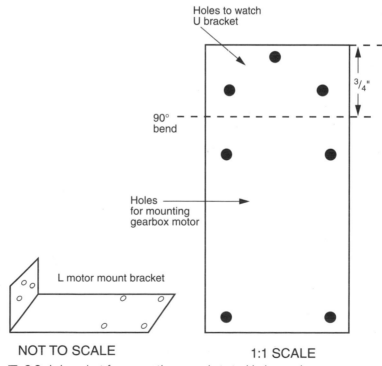

Holes to watch
U bracket

$\frac{3}{4}$"

90°
bend

Holes
for mounting
gearbox motor

L motor mount bracket

NOT TO SCALE

1:1 SCALE

■ **8.3** *L bracket for mounting gearbox to U channel*

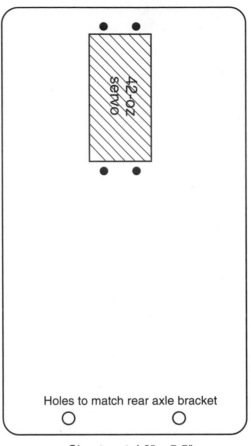

Holes to match rear axle bracket

Sheet metal 3" x 5.5"

■ **8.4** *Robot base showing cutout for 42-oz servo motor and holes for rear axle bracket*

fore bending it into shape. For the rear axle, I used the wire from a metal coat hanger.

To continue we need to mount the front drive wheel to the gearbox motor. The rubber wheel used in this prototype is made to friction fit a 3-millimeter (mm) (0.118″) shaft. The shaft diameter of the 100:1 gearbox motor is about 2 mm (0.078″).

To solve this size problem, I placed a 3″ long length of 3-mm hollow metal tubing onto the shaft of the gearbox motor. I used a flat-head screwdriver and hammer to secure the 3-mm tubing to the 2-mm shaft. First place the motor's shaft and tubing onto a hard (metal) surface that allows you to place force directly onto the shaft without causing any strain on the gears or motor. Next place the screwdriver head on the shaft-tubing assembly and hit it sharply with the

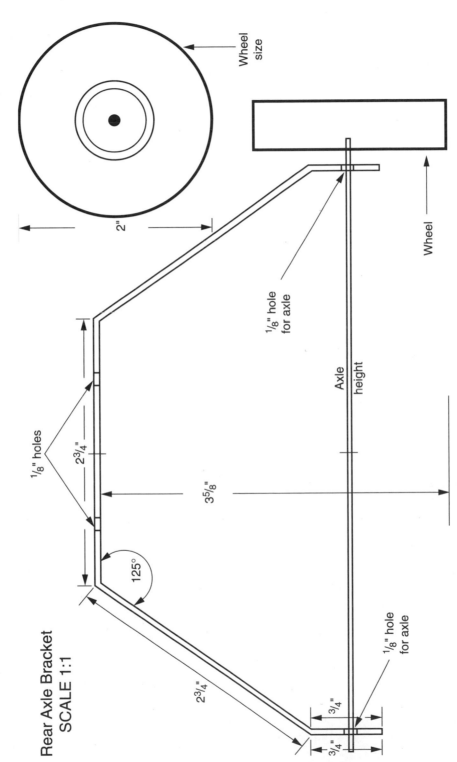

Rear Axle Bracket
SCALE 1:1

Wheel size

2"

$^1/_8$" hole
for axle

Axle
height

$^1/_8$" holes

$2^3/_4$"

$3^5/_8$"

125°

Wheel

$2^3/_4$"

$^1/_8$" hole
for axle

$^3/_4$"

$^3/_4$"

■ **8.5** *Rear axle bracket detail*

173

hammer. This force will cause the tubing to collapse onto the shaft making a strong friction fit. Strike the 3-mm tubing in one or two locations for insurance.

If one looks closely at the gearbox motor shaft, there is a keyway (flattened cutaway on the shaft) cut into the shaft. If you properly strike the tubing at that location to collapse the tubing into the keyway, you will create a very secure fitting between the motor shaft and tubing.

The drive wheel is mounted by pushing it onto the 3-mm tubing. The friction fit of the wheel is strong enough to drive the robot without any slippage. If you wish to mount the wheel permanently (something I have not done) to the shaft, try mixing slow-setting epoxy glue and coating the 3-mm shaft with it before mounting the wheel onto it.

### Counterbalance

When the gearbox motor is mounted on the U channel, the weight of the gearbox motor on one side makes the assembly unbalanced. To balance the U channel, I placed 3 to 4 oz of lead on the opposite side. I have $1/8''$ thick lead sheets lying around that I use to store radioactive isotopes. Cutting and drilling the lead is easy. You can mount any heavy object onto the shaft as a counterweight (like fender washers).

### Shell

The original tortoise robot had a transparent plastic shell. The shell was connected to a bump switch that caused the robot to go into "avoid" mode when activated. I looked at, tried, and rejected a number of different shells. Finally I was left with no choice other than to fabricate my own shell.

Rather than fabricate an entire shell, I made a bumper that encompasses the robot. The bumper is fabricated from $1/8'' \times 1/2'' \times 32''$ aluminum bar (see Fig. 8.6). The aluminum bar is marked at the center. Each bend required in the bumper is also marked in pencil. The material is placed in a vise at each pencil mark and bent to the angle required. The two ends of the aluminum bar end up at the center back of the bumper. These two ends are joined together using a $1/8'' \times 1/2'' \times 1''$ long piece of aluminum bar. A $1/8''$ hole is drilled on each end of the aluminum bar. Matching holes are drilled in the ends of the bumper. The bar is secured to the bumper using two 5-40 machine screws and nuts (see Fig. 8.7).

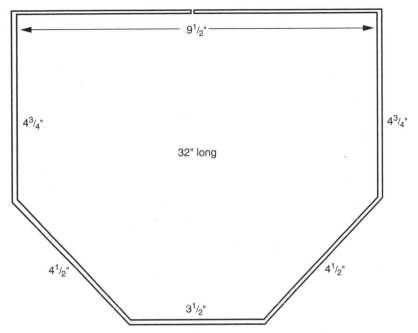

■ **8.6** *Top dimensional view of bumper fabricated from* ¹/₈″ × ¹/₂″ × *32″ aluminum bar*

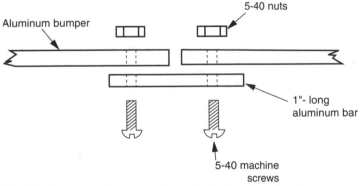

■ **8.7** *Cutaway close-up of aluminum bracket used to secure the open ends of the bumpers*

The upper bracket used to connect the bumper to the robot is identical to the front end of the bumper (see Fig. 8.8). The upper bracket is made from ¹/₈″ × ¹/₂″ × 14.5″ aluminum bar. As with the bumper, the center of the bar is marked and each bend required is also marked in pencil. The material is bent in the vise the same way as the bumper.

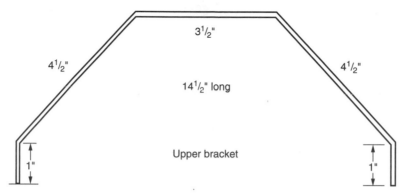

**■ 8.8** *Side dimensional view of upper bracket fabricated from ¹/₈″ ×*
*¹/₂″ × 14¹/₂″ aluminum bar*

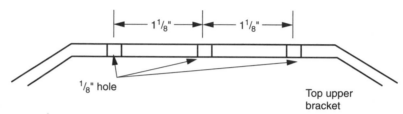

**■ 8.9** *Side dimensional view for hole placement in top of the upper*
*bracket*

### Finding the center of gravity

It is important to find the center-of-gravity line of the bumper, because this will mark the optimum location where the upper bracket should be attached. Rest the bumper on a length of aluminum bar. Move the bumper back and forth until it balances evenly on the aluminum bar. Mark the centerline positions on each side of the bumper. Drill a ¹/₈″ hole on each side. Drill matching holes on the ends of the upper bracket. Then secure the upper bracket to the bumper using 5-40 machine screws and nuts.

### Attaching bumper to robot base

The bumper is attached to the robot body by the upper bracket. Drill three ¹/₈″ holes in the top of the upper bracket. One ¹/₈″ hole is in the center and the two other holes are 1¹/₈″ away from the center hole (see Fig. 8.9). Three matching holes are drilled in the robot base behind the servo motor. The holes should be placed so that the bumper (once secured to the base) has adequate clearance (¹/₈″ to ¹/₄″) from the back wheels. The matching center hole on the base must be offset by moving the drilled hole forward on the base by about ¹/₄″.

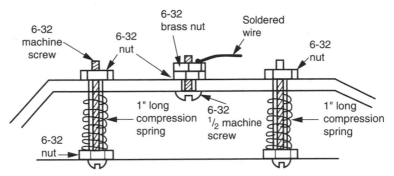

**■ 8.10** *Side view of upper bracket detailing the mounting of the upper bracket to the robot base using machine screws and compression springs. Also details bracket half of the tilt switch*

The bracket is secured to the base using two 1" long 6-32 machine screws; four 6-32 nuts; and two 1" long, 2-pound (lb) compression springs, with a $\frac{1}{8}$" center diameter (see Fig. 8.10). The tension and resiliency of the bumper can be adjusted by tightening or releasing the upper 6-32 machine screw nuts. Once assembled, the bumper will tilt back and close the bumper switch when the robot (bumper) encounters (pushes against) an obstacle.

### Bumper switch

The bumper switch makes use of the center holes. Looking back at Fig. 8.10, the center hole is fitted with a 6-32 machine screw held on by a standard (zinc-plated) nut, followed by a brass nut. The brass nut has a wire soldered to it. The purpose of this little assembly is just to attach a wire to the bracket-bumper assembly. Brass nuts are used because it is possible to solder wires to brass to make electrical connections. This is in contrast to the standard zinc-plated steel nuts that are very difficult (impossible) to solder.

The second half of the tile switch is comprised of a 1" 6-32 plastic machine screw and three 6-32 machine screw nuts, one of which must be brass with a wire soldered to it (see Fig. 8.11). Figure 8.12 is a close-up of the finished tilt switch. The assembly is adjusted so that the brass nut on the top of the 6-32 machine screw lies just underneath the upper aluminum bracket without touching. When the upper bracket tilts forward, contact is made between the aluminum bracket and brass nut, which is read as a switch closure.

### Photoresistor

The cadmium sulfide (CdS) photoresistors used in my prototype have a dark resistance of about 100K ohms and a light resistance of 10K ohms. The top of the 100:1 gearbox motor bracket is a perfect

*177*

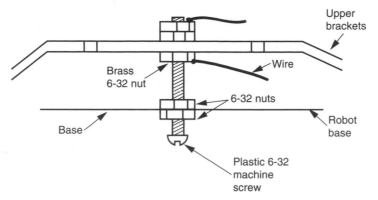

■ **8.11** *Side dimensional detail (robot base side of the tilt switch) of plastic screw with top brass nut*

■ **8.12** *Close-up photograph, detailing tilt switch and spring mounting of upper bracket*

shelf for mounting the photoresistor (see Fig. 8.13). I used a small piece of plastic to mount the photoresistors at a 45 degree angle up with an opaque vane mounted in between the photoresistors (see Fig. 8.14). Mounting the photoresistors on the drive wheel assembly keeps the sensors pointing in the same direction as the drive wheel. This replicates the function of the original tortoise robots.

Using two CdS photosensors in this configuration alleviates much of the computation needed to track a light source. This is the same

**■ 8.13** *Close-up photograph detailing front drive wheel, showing counterweight, drive wheel, gearbox motor, and light sensor with shroud*

operation as described in Chap. 6 for the light tracker circuit. The operation of the sensor array is shown in Fig. 8.15. When both sensors are equally illuminated, their respective resistances are approximately the same. As long as each sensor is within ±10 points of the other, the PIC program will see them as equal and doesn't move the servo motor (steering). When either sensor falls in the shadow of the main light source, the resistance variance between the sensors increases beyond the ±10-point range. The PIC microcontroller activates the servo motor to bring both sensors back under even illumination. In doing so, this steers the robot straight to the light source. If the sensors detect too great of a light intensity, the robot will go into avoid mode.

### Schematic

The schematic for the robot is shown in Fig. 8.16. Intelligence for the robot is provided by two PIC16F84 microcontrollers. The steering servo motor control signal is provided by RB3 off the PIC microcontroller number 2. The 100:1 gearbox motor is attached to an H-bridge consisting of components Q1 to Q4, D1 to D4, and R1 to R4. The H-bridge is controlled by the PIC microcontrollers RB1 and RB2 input/output (I/O) lines. Sensor readings of the CdS cell are read off pin RB4. RB5 reads the tilt switch to check if the robot has encountered an obstacle. I assembled the entire circuit on two

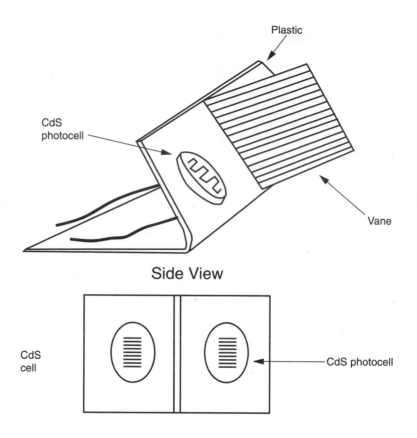

Side View

Front View

CdS cell

CdS photocell

■ **8.14** *Isometric view of sensory array*

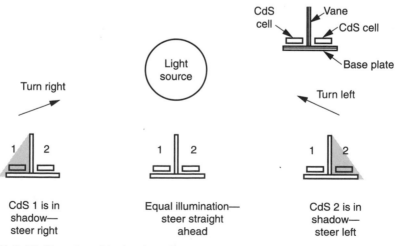

CdS cell

Vane

CdS cell

Base plate

Light source

Turn right

Turn left

CdS 1 is in shadow—steer right

Equal illumination—steer straight ahead

CdS 2 is in shadow—steer left

■ **8.15** *Functional behavior of sensor array*

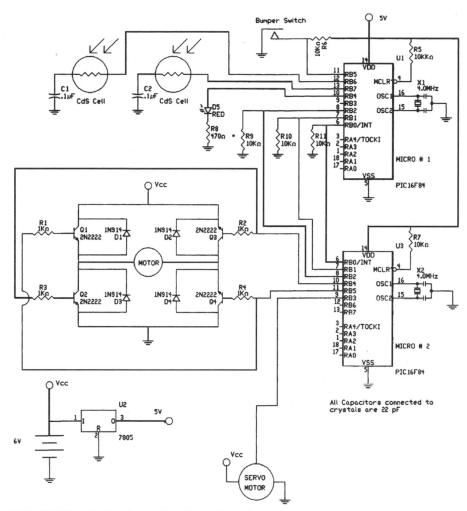

■ **8.16** *Electrical schematic of tortoise robot*

small solderless breadboards. The breadboards were mounted on the robot base, on top of the battery power pack.

To meet the original design expectations (accurately modeling the original Walter's tortoise robot for one), two microcontrollers are required. Distributing the computational workload between two processors produces much smoother operation from the robot.

The main reason a second microcontroller is needed is for the steering servo motor. A single microcontroller cannot read the two CdS photoresistors and accurately control (steer) the servo motor. Had I chosen a gearbox motor for steering the robot instead, using a single microcontroller would not have been a problem. On the bright side, the advantage to circumventing this problem is building a robot with two processors that operate in tandem (distributed processing).

I will assign the targeting of the light source and bumper switch detection to one microcontroller, called microcontroller 1. Controlling the steering and drive motor will be assigned to the second microcontroller, called microcontroller 2. To make this scheme work, it is necessary to have the microcontrollers communicate with one another. However, bidirectional communication isn't required; we will have one microcontroller talk and the other one listen.

**Microcontroller 1**    Microcontroller 1 will handle reading the CdS cells and bumper switch detection. It will communicate to microcontroller 2 using three I/O lines.

☐ I/O line 1 will communicate the status of CdS 1. If light falling on CdS 1 is brighter than light falling on CdS 2, then bring the line low. If equal, bring the line high.

☐ I/O line 2 will communicate the status of CdS 2. If light falling on CdS 2 is brighter than light falling on CdS 1, then bring the I/O line low. If equal, bring the line high.

☐ I/O line 3 will communicate either the status of the bumper switch or that the CdS cells are receiving too much light. In either case this will bring line 3 high.

**Microcontroller 2**    Microcontroller 2 will check the status of the three I/O lines and, based on the status, it will steer and move the robot as follows:

| CdS 1<br>Line 1 | CdS 2<br>Line 2 | Tilt switch<br>Line 3 | Results |
|:---:|:---:|:---:|---|
| 0 | 0 | 0 | Sleep mode, not moving |
| 1 | 1 | 0 | Move forward |
| 1 | 0 | 0 | Steer right, move forward |
| 0 | 1 | 0 | Steer left, move forward |
| X | X | 1 | Avoid mode |

X = don't care.

Accordingly, lines 1 and 2 represent the two CdS cells and line 3, the status of the bump switch.

### Adding sleep mode

I added a sleep mode for when the ambient light is very low. The robot moves forward when both CdS sensors receive approximately the same light intensity. The robot steers right or left when one CdS cell receives more light than the other. If each CdS cell receives too much light or the bump switch is activated, the robot goes into avoid mode.

### Power

Six-volt electrical power for the robot is supplied by a battery pack of four AA batteries. While I used this power supply for testing robot function, I suspect the batteries may wear out quickly.

## Program

The program flowchart is shown in Fig. 8.17. Upon power-up, the drive motor is off and the microcontroller begins scanning for the brightest light source using the servo motor. If a light source is too bright, the robot jumps into avoid mode. In avoid mode the robot backs away from the light source by reversing the drive motor while steering the drive wheel left or right. If the light isn't bright enough to activate the avoid mode, the robot steers in the direction of the light and activates the drive wheel forward. If the bumper switch is activated, the robot assumes it has hit an obstacle and the robot goes into avoid mode. But if the tilt switch is not activated (no collision), the program jumps to the beginning and the process continues scanning and moving to the brightest light source.

The program is written for the PICBASIC compiler that is directly programmed into a PIC16F84. The program should be able to be compiled and run with little or no modification on the PICBASIC Pro

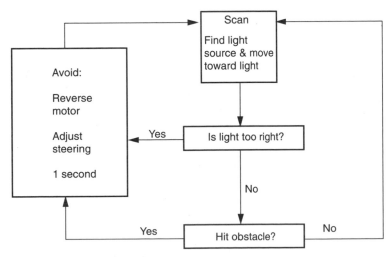

■ **8.17** *Program flowchart*

version. In-group variances in CdS sensors, drive motors, robot structure, and the like can be adjusted for or modified in the program.

## Program 1

```
'Microcontroller 1
start:
high 4:low 4    'Blink LED
b7 = 0
button 5,0,255,0,b7,1,avoid    'Check for obstacle
pot 7, 255, b0    'Read CdS cell 1
pot 6, 255, b1    'Read CdS cell 2
If b0 <= 250 then skip    'Is it dark enough to sleep?
If b1 >= 250 then slp    'Yes
skip:    'No
if b0 > 25 then skip2    'Is it too bright to live?
if b1 < 25 then avoid    'Yes
skip2:    'No
if b0 = b1 then straight    'Light is equal; go straight
if b0 > b1 then greater    'Check light intensity
if b0 < b1 then lesser    'Check light intensity
straight:
high 0: high 1: low 2    'Communicate to microcontroller 2
goto start    'To go straight
greater:
b2 = b0 − b1    'Check numerical difference
if b2 > 10 then rt    'If more than 10, turn right
goto straight    'If not, go straight
lesser:
b2 = b1 − b0    'Check numerical difference
if b2 > 10 then lt    'If more than 10, turn left
goto straight    'If not, go straight
rt:    'Turn right, send
high 0: low 1: low 2    'Communication to microcontroller 2
goto start
lt:    'Turn left, send
low 0:high 1: low 2    'Communication to microcontroller 2
goto start
Slp:    'Go asleep, send
low 0: low 1: low 2    'Communication to microcontroller 2
goto start
avoid:    'Avoid mode, send
low 0:low 1: high 2    'Communication to microcontroller 2
goto start
```

## Program 2

```
'Microcontroller 2
b4 = 150    'Initialize servo to midposition
start:
peek 6, b1   'Read communication from microcontroller 1
let b0 = b1 & 7   'Mask out except first 3 bits
if b0 = 0 then slp   'Time to sleep
if b0 = 1 then rt    'Turn right
if b0 = 2 then lt    'Turn left
if b0 = 3 then fw    'Move forward
if b0 = 4 then avoid    'Avoid mode
goto start
slp:
low 4: low 5    'Turn off motor
pulsout 3, b4    'Maintain servo motor
pause 18    'Timing for servo motor
goto start    'Read microcontroller 1
rt:    'Turn right
high 4: low 5   'Move forward
      if b4 > 200 then rt1:    'Is servo at maximum right?
      b4 = b4 + 1   'No
      rt1:    'Yes
pulsout 3, b4   'Turn servo
pause 18   'Adjust timing (55 Hz)
goto start    'Read microcontroller 1
lt:    'Turn left
high 4: low 5   'Move forward
      if b4 < 100 then lt1:    'Is servo at maximum left?
      b4 = b4 −1   'No
      lt1:    'Yes
pulsout 3, b4   'Turn servo
pause 18    'Adjust timing (55 Hz)
goto start    'Read microcontroller 1
fw:    'Forward
high 4: low 5   'Move forward
pulsout 3, b4   'Turn servo
pause 18    'Adjust timing (55 Hz)
goto start    'Read microcontroller 1
avoid:
low 4: high 5    'Move backward
if b4 > 150 then vr    'Check steering, veer right?
if b4 <= 150 then vl    'Check steering, veer left?
vr:    'Veer right
```

*Behavioral-based robotics, neural networks, nervous nets, and subsumption architecture*

```
b5 = b4 - 30    'Create servo direction
    for b6 = 1 to 120    'Make 2-s timing loop
    pulsout 3, b5    'Turn servo
    pause 18    'Adjust timing (55 Hz)
    next b6    'Loop
goto start    'Read microcontroller 1
vl:    'Veer left
b5 = b4 + 30    'Create servo direction
    for b6 = 1 to 120    'Make 2-s timing loop
    pulsout 3, b5    'Turn servo
    pause 18    'Adjust timing (55 Hz)
    next b6    'Loop
goto start
```

The finished robot is shown in Fig. 8.18.

## Behavior

The robot needs to function in a low-light environment, where it can clearly see a bright light source. The light level required of my robot was so low I needed to fabricate tiny sunglasses out of colored plastic to reduce the light intensity hitting the CdS photocells.

The prototype robot exhibits the following behavior. In ambient light (no bright light source) the robot travels in a straight line (or circle depending upon the last light source target). If the ambient

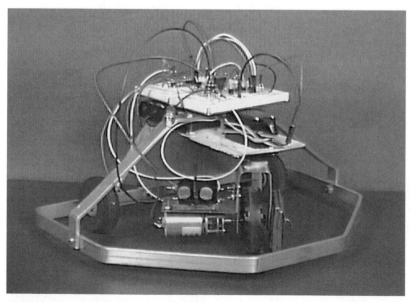

■ **8.18** *Front view of finished robot*

light is too bright, it jerks backward. With a mediocre light source, it will aim and travel toward the light.

The program can be developed further to explore more interesting and exotic behaviors. Before doing so, let's first look at how the standard program functions. Program 1 for microcontroller 1 primarily checks the sensors and transmits the results to microcontroller 2. In this program you can modify the sensor sensitivity to compensate for different sensors, for instance, by using the lines

```
if b0 <= 250 then skip    'Is it dark enough to sleep?
if b1 >= 250 then slp    'Yes
skip:    'No
```

The maximum reading from the sensor can be 255 (total darkness). This may be raised to increase the ambient light intensity for sleep.

The brightness that triggers the avoid mode may be modified by using the following lines:

```
If b0 > 25 then skip2    'Is it too bright to live?
If b1 < 25 then avoid    'Yes
skip2:    'No
```

Increasing the numerical value, in this case 25, decreases the light intensity that puts the robot into avoid mode. Decreasing the numerical value increases the light intensity needed to throw the robot into avoid mode. In most cases you will want to *decrease* this number. However, I would advise not going below a numerical value of 9, because even at full light saturation of the CdS cell, its resistance never drops to zero. And in my light saturation tests the sensor never yielded a value less than 5.

Tolerance between the two CdS photoresistors may be increased or decreased by modifying the numerical allowable difference in subroutines greater and lesser.

```
greater:
b2 = b0 - b1
if b2 > 10 then rt
goto straight
lesser:
b2 = b1 - b0
if b2 > 10 then lt
goto straight
```

In addition, one could create handedness in the robot (right- or left-handed) by modifying either the greater or lesser subroutine, but not both. This will create a robot that is more likely to turn in one direction than the other. For instance, if we modified the line if b2 > 10 then lt in the lesser subroutine to read if b2 > 15 then lt, we would create a robot that is more likely to turn to the right.

This robot offers many opportunities to robotists and experimenters for continued experimentation and development both in hardware and software.

## Parts list for the Walter tortoise robot

- ☐ (1) 12″ × 12″ sheet metal, 22 or 24 gauge
- ☐ (1) $\frac{1}{8}$″ × $\frac{1}{2}$″ × 12″ long aluminum bar
- ☐ (1) 42-oz torque hobby servo motor
- ☐ (1) 100:1 gearbox motor (or similar)
- ☐ 3-48 machine screws and nuts
- ☐ 0-80 machine screws and nuts
- ☐ (1) $\frac{1}{8}$″ × $\frac{1}{2}$″ × 32″ long aluminum bar
- ☐ (1) $\frac{1}{8}$″ × $\frac{1}{2}$″ × 14$\frac{1}{2}$″ long aluminum bar
- ☐ (1) $\frac{1}{8}$″ × $\frac{1}{2}$″ × 2″ long aluminum bar
- ☐ (1) 42-oz torque standard servo motor
- ☐ (1) 100:1 gearbox DC motor
- ☐ (1) 2″ diameter drive wheel—friction fit to 3-mm shaft
- ☐ (1) 2-mm ID, 3-mm OD steel or brass tubing
- ☐ (2) CdS photocells, 100K-ohm dark, 10K-ohm light
- ☐ [4 (Q1-Q4)] 2N2222 NPN transistors
- ☐ [4 (D1-D4)] 1N914 diodes
- ☐ (1 D5) Red LED
- ☐ [4 (R1-R4)] 1K-ohm, $\frac{1}{4}$-W resistors
- ☐ [6 (R5-R7, R9-R11)] 10K-ohm, $\frac{1}{4}$-W resistors
- ☐ [1 (R8)] 470-ohm, $\frac{1}{4}$-W resistor
- ☐ (4) 22-pF caps
- ☐ [2 (C1,C2)] 0.1-μF capacitors
- ☐ [2 (X1, X2)] 4-MHz crystal
- ☐ [1 (Q5)] 7805 voltage regulator
- ☐ [2 (IC1, IC2)] 16F84-04 PIC microcontroller
- ☐ Miscellaneous: 5-40 machine screw and nuts, plastic 6-32 × 1″ machine screws, 6-32 brass nuts, 1″ long compression springs (2 lb)

## Suppliers

- ☐ Aluminum bars, machine screws, tubing, and compression springs are available in most well-stocked hardware stores.
- ☐ Servo motors may be purchased at hobby shops or electronic distributors.
- ☐ Electronic components may be purchased from Radio Shack, Images Company, Jameco Electronics, or JDR Electronics.
- ☐ PIC microcontroller and front drive wheel may be purchased at Images Company.

Images Company
39 Seneca Loop
Staten Island, NY 10314
(718) 698-8305

Jameco
1355 Shoreway Rd.
Belmont, CA 94002
(650) 592-8097

JDR
1850 South 10 St.
San Jose, CA 95112
(800) 538-5005

## Building an intelligent photovore robot

Let's see if we can create what may appear as intelligent behavior in a photovore robot. In Chap. 6 we programmed a photoresistive light tracking system. The tracking system locked onto a light source and tracked it. When we placed the same tracking system on a copy of Walter's tortoise robot, it directed the robot toward a light source. This steering behavior can be considered the first stimulus-response layer.

The program illustrates how the rule-based microcontrollers can simulate neural functions. For the sake of an example, let's now design a neural circuit that performs the same function without any rule-based intelligence.

Figure 8.19 uses an 8-pin dip single-voltage-supply dual operational amplifier (op-amp). The op-amps are configured as comparators. Comparators were covered more fully in Chap. 5. If you have any questions about Fig. 8.19, review Chap. 5. Two CdS photoresistors are wired in series forming a voltage divider. The output of the

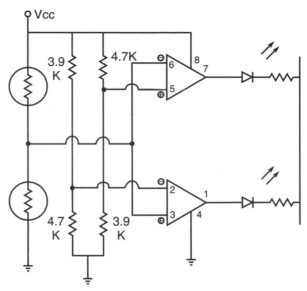

■ **8.19** *Dual op-amp neural comparator circuit*

photoresistor voltage divider is fed into the inverting input of one op-amp and the noninverting input of the other op-amp.

Two other voltage dividers are needed. They are symmetric but opposite. One voltage divider has a 3.9K-ohm resistor connected to Vcc and a 4.7K-ohm resistor connected to ground. The second voltage divider uses the same value resistors, in opposite positions.

When both photoresistors are evenly illuminated, neither LED is lit. Cover one or the other photoresistor and the corresponding LED will light.

Each op-amp acts like a simple electronic neuron. When the electrical stimulus falls above or below (depending upon which op-amp we're talking about) a threshold (determined by the 3.9K-ohm and 4.7K-ohm voltage dividers), the neuron fires. The firing of the neuron (or outputs of the op-amp) can be used to turn on a DC motor using an NPN transistor (see Fig. 8.20). The DC motors in turn provide movement and direction for the photovore robot.

To create a simple photovore robot, a chassis is designed that has two gearbox DC motors (see Fig. 8.21). When both motors are powered, the robot moves forward in a straight line. When one motor is turned off, the motor that still receives power will turn the robot left or right.

For our photovore robot, we need both motors to be powered when the two photoresistors are evenly illuminated. Running the outputs of each op-amp into an inverting buffer located just before the NPN transistor accomplishes this task (see Fig. 8.22).

## Behavior

When one photoresistor receives less light than the other, the corresponding motor will turn off, allowing the motor that's still powered to turn the robot toward the light source. When the robot turns so that both photoresistors are again evenly illuminated, both motors turn on, allowing the robot to travel toward the light source.

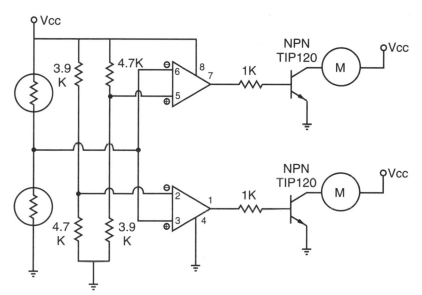

■ **8.20** *Neural comparator DC motor control circuit*

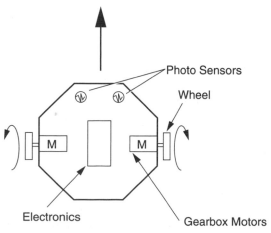

■ **8.21** *Outline diagram of photovore robot*

*Behavioral-based robotics, neural networks, nervous nets, and subsumption architecture*

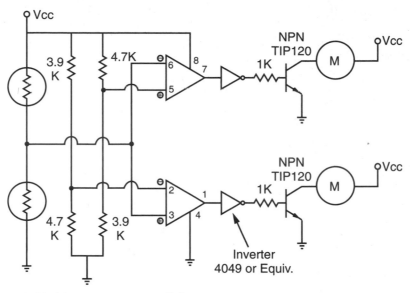

■ **8.22** *Neural comparator DC motor control circuit with inverters*

### Light avoidance

If we crisscross the outputs of the op-amps to the motors, the behavior reverses. Instead of moving toward a light source the robot now avoids light and seeks shelter.

## Adding behavior (feeding)

We can add behavior to the photovore by adding another stimulus-response layer (see Fig. 8.23). This will be another light-activated comparator circuit that facilitates feeding. Comparators were covered more fully in Chap. 5. If you have any questions about Fig. 8.23, go back to that chapter. The second layer is placed on top of the first layer. When the light intensity is great enough, this threshold detector cuts power to the first layer and the motor drive system. If we place a number of photovoltaic cells and a diode, the electric power generated from the photovoltaics can trickle charge a nickel-cadmium (NiCd) power pack (Vcc). Let's call this function "feeding."

## Still more behavior (resting)

We don't want our photovore traveling around in the dark wasting precious energy. So let's add another layer. The third layer is another light threshold detector (see Fig. 8.24). This detector cuts power to the first layer, motor drive system, and second layer in darkness

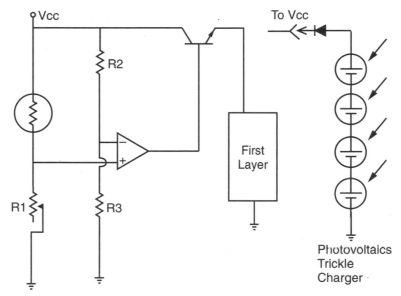

■ **8.23** *Feeding behavior comparator circuit*

or near darkness. When a sufficient amount of ambient light is reintroduced, power to the first layer, drive system, and second layer is restored.

## Emergent behavior

Let's look at the behavior of our three-layer stimulus-response photovore robot and see if we can classify its behavior as intelligent. In complete darkness the robot remains still, conserving all its power via layer 3. As ambient light is introduced and increased, layer 3 restores power to the drive system and first two layers. At this point, layer 1 takes over and controls the direction of the robot. The robot searches and moves toward the source of light. As the robot moves toward the light source, the light intensity increases. When the light reaches a sufficient intensity, layer 2 cuts power to the drive system allowing the robot to feed (charge its batteries) through the photovoltaics.

Whether you decide to classify this robotic behavior as intelligent or not is an individual preference and one that can clearly be debated on both sides of the fence. In the least, it illustrates how complex behavior patterns can be generated using a layered stimulus response.

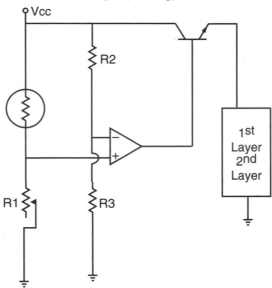

**3rd Layer (Resting)**

■ **8.24** *Resting behavior comparator circuit*

# BEAM robotics

Mark Tilden founded BEAM robotics while at the University of Waterloo in Canada. The inspiration for BEAM-style robots came from a talk given by Rodney Brooks of MIT that Mark had attended in 1989. Dr. Rodney Brooks's approach to robotics is a stimulus-response system he refers to as "subsumption architecture."

*BEAM* is a multidimensional acronym that loosely stands for biology, electronics, aesthetics, and mechanics. I say "loosely" because there are numerous groups of words that can be and are at times substituted in the acronym, for instance biotechnology, evolution, analog, and modularity.

## BEAM competition

There is an annual Olympic competition for BEAM robotics with 14 events. The first BEAM competition was held in 1991. The inspiration for the BEAM games came from the first international Robot Olympics held in Glasgow, Scotland, in 1990. A central idea to the BEAM philosophy is robotic evolution. Start simple and evolve toward complex systems. As illustrated, the idea is to break away from standard robotic design, using top-heavy CPUs for control, and embrace a bottom-up approach using a layered stimulus response (neural network, nervous network systems). Mark Tilden calls his stimulus-response mechanisms "nervous nets."

Tilden has designed a number of interesting robots (see Fig. 8.25). They employ nervous net systems that are made using transistors. Since the nervous net system is patented (by Mark Tilden) and unpublished schematics for his nervous net system are not readily available, I do not have any nervous net system schematics to present. However, Tilden has a book in the works titled *Living Machines*.

Figure 8.26 is titled Gumby Trks. This is a type of biomechanical walker that is being designed for a variety of terrains. Here Gumby 1.0, an eight-transistor imbedded-bicore walker about 1 ft long, makes tracks across a sand desert.

Figure 8.27 is titled Walkman 1.0. The first of the 12-transistor "Microcore" walkers, this device was put together from the remains of five similar Walkman cassette players. It has seven sensors including two eyes and can handle very complex terrains with its five-motor design.

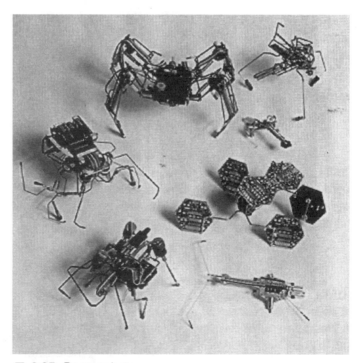

■ **8.25** *Beam robots*

■ **8.26** *Gumby Trks*

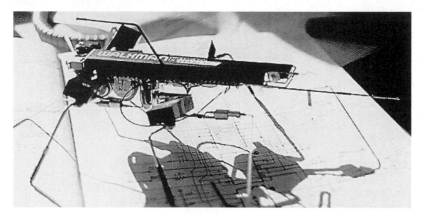

■ **8.27** *Walkman 1.0*

### Electronic flotsam

BEAM robotists pride themselves on using discarded electronics in the construction of their robots: for instance, solar cells from calculators, high-efficiency electric motors from Walkmans and cassette players, along with pulleys, switches, capacitors, components, gears, and solenoids. Gathering this electronic flotsam and converting it into useful robots is a project in recycling engineering.

The BEAM robotic competition is open to everyone. All competitors start on equal footing. Seven-year-old robotists have as much chance of winning as a professor from a prestigious college. In some cases the 7-year-old won!

### Competitions

The following is a brief synopsis of the competitions held at the BEAM games. Complete descriptions of the events and rules can be found in the BEAM guide available from the University of California. The address is listed at the end of this chapter.

### Solaroller

Create a solar-powered robot racer that fits into a 6″ cube. Maximum solar cell size is $\frac{1}{2}″ \times 2\frac{1}{2}″$ [1.25 square inches (in$^2$)]. Track length is 1 meter (m); width is 6″. Competitors race in full sunlight (or 500-W halogen lamp equivalent).

☐ *Class A.*   Race on level sheet of glass.

☐ *Class B.*   Race on rough terrain.

### Photovore

Create a solar-powered, goal-seeking robot that can fit into a 7″ cube. The robot will be placed with other competitors in a closed "Jurassic Park" for 30 hours. Those robots that show the best survival, exploration, confrontation, speed, and power efficiency, determined by review of photos and video, will be the winners.

### Aquavore

Create a solar-powered robot that can fit inside a 7″ cube and be able to swim the length of a 55-gallon fish tank (distance is approximately 1 m). A 6″ high wall will be placed halfway in the tank that the competitor must pass to reach the finish line.

### Robot limbo

Create a robot that can fit inside a 7″ cube that will run through a simple maze. Solar power is not required for this competition, but is recommended.

### Robot rope climbing

Create a robot that can climb up a meter of rope and then back down. The fastest robot wins. The rope is 40-lb test nylon fishing line. The robot must fit in a 20″ cube.

### Robot high jump and long jump

**A class**   Create a robot that can jump with its entire mass into the air three times using the power from one optional battery. Robot must fit in a 1-square-foot (ft$^2$) space.

**B class**   Create a robot that can jump with its entire mass forward three times using the power from one optional battery. Robot must fit within a 1-ft$^2$ space.

### Legged robots

Legged robots compete with each other. Robots are given points based upon their capabilities to walk over various terrains and negotiate obstacles. No size restriction.

### Innovation machines

Create a new device, the purpose of which need not be obvious. Competitors are judged on quality of the design and assembly, broadness of scope, and weirdness of application.

### Robot art/best modified appliance competition

Create a robot that can draw or generate art. The generation of art may be the movement of the robot itself. An example given is a solar flower that opens slowly and snaps closed when light shines upon it.

**Class A**  Robots built completely from scratch.

**Class B**  Modified devices, toys, appliances, etc.

### Robot sumo wrestling

**Class A**  Robots are paired together in competition. Each robot attempts to push the other off the edge of a 5-ft-round platform. Robots can be self-contained, tethered, or radio controlled.

**Class B**  Robots try to push each other off a 6-ft-round platform.

### Nanomouse competition

Create a self-contained robotic mouse that can run through a maze. The robot's footprint must be no larger than 10 cm × 10 cm. No restriction on height.

### Micromouse competition

Create a self-contained robotic mouse that can run through a maze. The robot's footprint must be no larger than 25 cm × 25 cm. No restriction on height.

### Aerobot competition

Create a flying robot that will launch itself, fly into a 25-ft × 25-ft drop zone, find a randomly placed target in the drop zone, drop a marker on it, and then return to its launch pad.

### Miscellaneous competitions

If you have built a robot that doesn't fit in the outlined categories, it may be entered in the miscellaneous category.

## Getting the BEAM guide

Complete 120-page BEAM guides may be purchased for $20.00. Make checks payable to the University of California: BEAM Games. For current information call or write to:

BEAM Robot Olympics c/o Mark W. Tilden
Mail Stop D449
Los Alamos National Labs
Los Alamos, NM 87545
(505) 667-2902

The internet address for the BEAM games is

http://www.nis.lanl.gov/projects/robot/

## Join in

The BEAM competitions are open to all robotists. You can enter a robot in the competition or just attend the event for fun. Contact the BEAM Robotic Olympics, address given above for up-to-date information. The following internet site provides plans for building a simple solar roller robot:

http://www.imagesco.com

# Telepresence robot

# 9

IN THIS CHAPTER WE WILL BUILD A TELEPRESENCE ROBOT (T-bot). Telepresence robots are expanding into a variety of science, entertainment, business, military, exploratory, and industrial applications as was illustrated in Chap. 2.

## What's in a name?

The late science fiction writer Robert Heinlein is credited as the first person to predict the use of telepresence robots in his 1940 science fiction novel titled *Waldo.* In the story, a human operates mechanical puppets, called "waldos," to do his bidding from a remote location.

Rather than use the term "waldo," I found the word "golem" from Yiddish mythology more suitable. The story of the golem describes a human spirit who intentionally places itself in a clay figurine. The spirit controls the clay figurine, bidding it to do that which the spirit would not or could not do in its human form. Once the golem's work is finished, the spirit returns to its human form. This definition adequately describes the new science of telepresence. I therefore have named my telepresence robot Golem I.

## What is telepresence?

*Telepresence* is a high-fidelity form of remote control that attempts to project the senses of the human operator into a robot at a distant site. The feedback interfaces used to create a telepresence system are the same as used in virtual reality (VR). Figure 9.1 illustrates a basic telepresence system.

In virtual reality we achieve immersion into a synthetic computer-generated environment by fooling our senses, as best we can, to

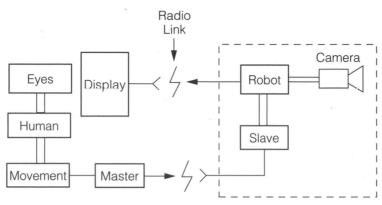

**9.1** *Outline of a basic telepresence system*

believe in and interact with the computer's synthesized environment. In telepresence the environment is real but remote. So, instead of a computer generating a synthetic environment, the sensors placed on the remote robot feed all spatial and environmental information to the user, in such a way that the user actually feels that he or she is there.

On this human side, as stated before, the same VR equipment is used to provide sufficient information from the remote sensors to fool our senses into believing that the environment is real and present. Different levels of presence are achieved depending upon the fidelity of the interfacing devices. A humanoid robot that could accurately follow human movement, gestures, locomotion, and balance while providing visual, thermal, tactile, and force reflection over its entire exoskeleton to the human operator would be a perfect golem. The illusion created is that the operator has merged or is contained within the robot structure.

Current telepresence systems fall quite short of this goal. In many cases the remote robot is a vehicle, like the one we shall build. The best telepresence existence available using these rudimentary T-bots allows one to believe he or she is actually driving the vehicle from inside.

T-bots can be built to explore and operate in harsh or hazardous environments. A partial list of remote environments include arctic waters, ocean floors, forest fires, active volcanoes, nuclear reactors, the Moon, Mars, or anything in between.

## System substructure

The framework upon which we will build our T-bot is a radio-controlled (R/C) electric car. Ideally the model car should have proportional steering and speed control. This is the type of model used in building the prototype. A less-expensive R/C model may be used, but you won't have as much control when driving.

Figure 9.2 is a photograph of the R/C model car. It has a spring suspension system. The suspension system can be incorporated with a rumble-and-tilt sensor system to provide a feel of the terrain. But we are getting ahead of ourselves.

Purchase an R/C car that is bundled with a battery charger and rechargeable batteries. With some R/C models these items must be purchased separately.

## A little on R/C models

Radio-controlled models have evolved into a popular hobby. There are R/C airplanes, helicopters, gliders, powerboats, submarines, cars, motorcycles, etc. Most models are suitable shells and springboards for golem-type robots.

Not long ago, R/C models were exclusively gas powered. In the late 1970s, improvements in battery technology and electric motors made electric-powered vehicles a viable option.

Model R/C cars are typically controlled using a two-channel transmitter/receiver. One channel controls steering and the other channel controls the throttle. Each transmitter signal is controlled by a potentiometer inside the transmitter. The steering potentiometer is often connected to a small steering wheel on the transmitter control. The throttle is usually connected to a trigger or stick.

■ **9.2** *R/C model car used in telepresence system*

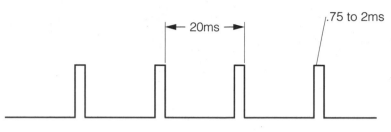

Pulse Signal to Servo Motor

■ **9.3** *Pulse width used to control a servo motor*

An encoder chip in the transmitter modulates the pulse width on the transmitter's carrier signal. The pulse width is based on the position (resistance) of the potentiometer's shaft. The pulse widths are varied between 1 and 2 milliseconds (ms) (see Fig. 9.3). When the potentiometer is in its center position, the pulse width corresponding to that channel is 1.5 ms. When the control is pushed to one extreme, the pulse width increases to 2 ms. When pushed to the opposite extreme, the pulse width shrinks to 1 ms.

The receiver decodes the pulses on the carrier signal and sends them to their respective servo motors. The servo motor is an integral unit, containing a motor, gearbox, output shaft, and a printed circuit board (PCB). The PCB on the inside of the servo motor generates a reference pulse that is based on the position of an internal potentiometer connected to the output shaft. A decoder chip on the internal PCB compares the incoming pulses from the receiver to the reference pulses. The servo motor attempts to match the pulse widths of the two signals by adjusting the position of the servo motor's output shaft. This is how the servo motor tracks and holds its position based on the signal from the transmitter.

## Eyes

The eye(s) for our T-bot is a miniature color video camera system with audio (see Fig. 9.4). The color camera system includes both a 2.4-gigahertz (GHz) transmitter and receiver. The camera system cost is approximately $99.95.

The overall size of the camera is small. It is mounted to the body of the transmitter by a small angled bracket. The video camera is small enough so that two video cameras are capable of being mounted side by side and have the approximate interpupilary distance (IPD) of 63 millimeters (mm) between lenses. Mounting a pair of cameras like this will enable the T-bot to transmit realistic stereo pictures to the operator. For the prototype we will use just one camera; later we will discuss improvements to the system that

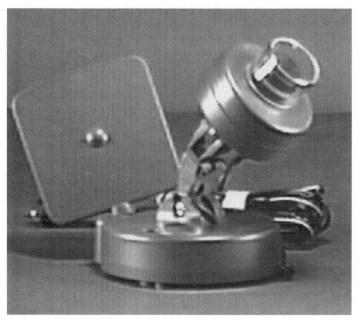

■ **9.4** *Color camera with audio and 2.4-GHz transmitter*

will add stereo-vision to provide depth perception to improve telepresence and operation.

The prototype T-bot contains a single miniature video camera with audio. The T-bot subsystems are built in modular form. Therefore, if the reader wishes to create a stereo system in the future for the T-bot, the video components are reusable.

## Construction

Construction of the T-bot begins by analyzing the chassis of the model car. Most R/C model cars have an outer cosmetic shell that makes the model look like a standard vehicle: car, truck, landrover, etc. Remove the outer cosmetic shell of the R/C model. Secure the equipment directly onto the chassis.

The Golem I requires a separate power supply for the miniature color camera (see Fig. 9.5). This 6-volt (6V) battery pack is made for use with the camera-transmitter pair. The battery pack should last approximately 4 to 6 hours (h) with four fresh AA batteries.

To keep the component mounting simple and modular, we will make liberal use of Velcro material. Velcro material is typically sold in strips, by foot increments. The Velcro strip is made of two mating strips of material that adhere to one another. Each strip

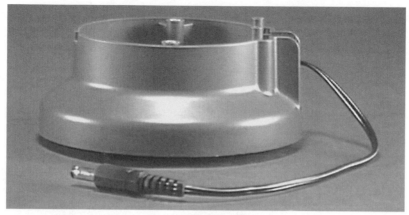

■ **9.5** *Six-volt battery pack holds 4 AA batteries*

■ **9.6** *Video camera system, camera transmitter, and receiver*

has self-adhesive on the back. The Velcro material is mounted to the chassis with the mating strip placed on the component to be mounted.

## 2.4-GHz video system

The 2.4-GHz transmitter is an integral part of the color camera (see Fig. 9.6). The 2.4-GHz receiver is a separate unit. The receiver has two RCA jacks, one for video out and the other for audio out. These are connected via RCA cables to the video in and audio in of a TV set, monitor, or video cassette recorder (VCR).

### Mounting the video camera

There are two options regarding the mounting of the video camera. The video camera and transmitter attached to the battery pack are shown in Fig. 9.7. You can start out with a fixed camera position, or the video camera can be mounted on top of a servo motor for camera movement and tracking. The servo motor mounting is naturally more complex and requires the building of a separate R/C control for panning the camera left and right. In a high-fidelity system, the camera panning would be linked to a head-tracking unit. So if the operator turned his or her head left or right, the camera on the mobile robot would also pan left or right in sync with the operator's head. To make this work properly, the operator should be wearing a VR-style head-mounted display (HMD). This is a lot of work. I advise using a fixed camera position first, to keep the construction simple.

### Fixed mounting

You need a small section of Velcro secured to the bottom of the camera's battery pack. The mating piece of Velcro is secured to the chassis of the R/C car. The finished Golem robot is illustrated in Fig. 9.8.

## Driving via telepresence

You can drive the car remotely using the radio controls while looking at the TV monitor. The camera is equipped with a microphone, so you can hear audio from around the R/C car when you are driving.

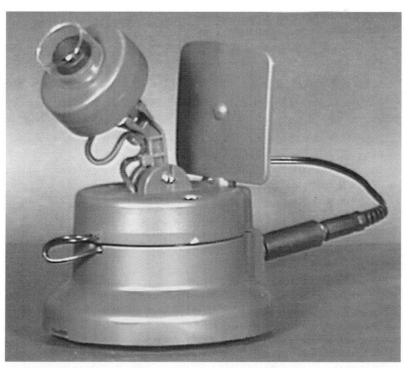

■ **9.7** *Color camera with battery pack ready for mounting on R/C car*

■ **9.8** *Color camera mounted on R/C car*

### Talk

You can purchase a pair of inexpensive children's walkie-talkies. Place one walkie-talkie on the T-bot. You can speak from the T-bot using the second walkie-talkie.

### Adding realistic car controls

Golem I uses the standard radio controls that came with the car. The realism of the telepresence system can be greatly improved by adding realistic car controls. This isn't difficult. It involves removing the circuit and potentiometers from the radio transmitter control and making a mock-up steering wheel for one potentiometer and a foot pedal throttle control for the other potentiometer.

## Improving the telepresence system

Golem I is a basic telepresence system that can be improved upon with a little thought. Improvements will add cost to the system. However, these subsystem improvements can be added over a time.

### Stereo-vision

Implementing high-fidelity stereo-vision on the Golem I is a worthwhile endeavor. There are great benefits to be derived from this experimentation, namely, depth perception. This is still an area where significant contributions can be made. Before undertaking this project it is important to realize that an HMD that supplies stereo images to the user will be needed to view stereo images transmitted from the T-bot.

The small size of the miniature video cameras is good for stereographic imaging. It allows two cameras to be positioned side by side at the same interocular distance (IOD) as that of human eyes. To be more specific, the average interocular distance (pupil to pupil) for adult humans is about 63 mm. The camera's lenses can be positioned, from center to center, this same distance apart to mimic the IOD humans use for depth perception. The transmitters for each camera must be tuned to transmit on different frequencies. This allows the HMD video receivers to accurately display the right eye image to the right eye and the left eye image to the left eye.

Seeing in stereo from the Golem provides the operator with depth perception when he or she is driving. Stereo-vision becomes increasingly more important when depth perception is needed, for instance when using a robotic arm. Being able to see the manipulator (robot arm) move along the Z axis in a three-dimensional (3D) coordinate system (X, Y, and Z) allows for efficient operation.

It is difficult for an operator to efficiently use robotic arms via telepresence when the Z dimension (depth) is lost in a monocular view. Operators are forced to gently bump into objects to approximate the robotic manipulator's location along the Z axis.

The same is true when driving a telepresence car. One quickly loses depth perception, and it becomes difficult to determine how far ahead of the vehicle something is.

When the stereo system is set up, the operator of the vehicle will see the T-bot's environment as a 3D picture. However, the stereo image transmitted will not contain the very important convergence clues. Much of our distance reckoning incorporates convergence clues we get from our eyes. *Convergence* is the angle our eyes rotate inward when viewing an object. An object very close to us will cause our eyes to rotate inward. In contrast, when viewing an object that is far away, our eyes look straight ahead. The brain automatically brings this convergence information to bear in calculating distance.

The stereo video cameras are in a fixed position looking straight ahead. To add convergence clues would require eye tracking engineering. A feedback-providing HMD would need to constantly ascertain the operator's eye convergence. The eye tracking (convergence) information would be transmitted to servo motors that hold the video cameras and would converge the video cameras in direct proportion to the operator's eye convergence.

This type of master-slave system, as far as I know, has not been built. It needs to be determined how accurate this system would be in helping an operator gauge distance. While building this system is beyond the scope of this book, it is not beyond the scope of an avid experimenter.

### Digital compass

Chapter 5 includes plans for a digital compass that is suitable for use with Golem. The compass can be set up in two different ways. The first method keeps the light-emitting diodes (LEDs) of the compass in the visual field of the video camera. A quick look informs the operator in which direction Golem is traveling. The second option uses a radio link between the digital output of the compass on the T-bot and the remote location of the operator.

### Rumble interface

When driving the model car via telepresence, you cannot feel the tilt or roughness of the road as you drive. To incorporate a rumble feature into the system, you could use the spring suspension of the model car. Any number of sensors can be used for this purpose, for instance, piezoelectric transducers, Hall devices, and strain gauges.

The challenge to the experimenter is not in detecting the rumble, but in providing that information to the seat of the operator. Most motion platforms use expensive pneumatic and hydraulic systems. If cost is an issue, this isn't an option.

A cheaper solution can be found in the ThunderSeat by Thunder-Seat Technologies. The Thunderseat utilizes any sound source to generate vibratory sensations. It contains a subwoofer speaker coupled to an acoustical wave chamber inside the seat. The wave chamber vibrates the entire seat. The low-frequency (woofer) speaker can handle up to 100 watts (W) of power. The frequency response of the system is 50 Hz to 3.7 kHz. Originally designed to work with flight simulator programs running with a sound card on a personal computer (PC), the output from the sound card is fed into an amplifier and then to the Thunderseat.

### Tilt interface

As with the rumble interface, there are several transducers one can use to determine tilt (see Chap. 5). One tilt sensor uses a steel ball in a plastic enclosure. When tilted, the steel ball makes contact with electrodes placed in the enclosure. Mercury switches may also be used.

Electrolytic tilt sensors are expensive, but are excellent sensors. A single electrolytic sensor can provide tilt information from two axes. The hermetically sealed sensor has one center electrode surrounded by four equidistance electrodes. As the electrolytic fluid makes contact with the internal electrodes, the alternating current (AC) resistance between the electrodes varies in proportion to the degree of tilt.

Unfortunately, the electrolytic sensors cannot be read using a direct current (DC) voltage source. This would cause the deposits to form on the electrodes, rendering them useless. Instead an AC voltage of approximately 3 V with a frequency of 1000 Hz is fed to the sensor. The AC voltage from the center electrode is in proportion to the tilt of the sensor.

If one were to use the electrolytic tilt sensor, I can suggest one way to set up the information flow. Connect the AC output of the tilt sensor to a bridge rectifier to obtain a DC equivalent voltage. The DC voltage is fed to a voltage-controlled oscillator (VCO). The VCO output frequency varies in proportion to the input voltage. The output of the VCO is transmitted over a radio link to a receiver on the motion platform. The receiver reads the frequency (tilt) and activates a proportional control to tilt the platform.

Spectron, Inc., offers an integrated circuit, the SA40011, that simplifies interfacing electrolytic tilt sensors. The DC output from the SA40011 can be fed to a VCO as described before.

Again, implementing tilt to the operator is the difficult part of the system. Proportional pneumatic or hydraulic systems can be employed to the seat to provide tilt.

## Greater video range

The video range of our small transmitter is approximately 100 to 300 feet (ft). Obviously for longer distances another system needs to be employed; it is called amateur television.

Amateur television (ATV) has been around for a number of years. It's been a method for radio amateurs to communicate via two-way television. ATV had been the province of the elite radio hobbyist due to the expensive cost of equipment. However, recent advances in solid-state technology have changed that.

The components for a 5-W ATV system can be purchased for $200, excluding TV monitor and video camera. The video cameras on Golem I are suitable for ATV use. Less-powerful ATV systems, $^3/_4$ W, can be purchased for under $100.

A Technician Class amateur license is required to operate these systems legally in the United States. Currently the Technician Class license no longer requires a knowledge of Morse code. Interested readers should contact a local amateur radio club for more information. Or you may write American Radio Relay League (ARRL), 225 Main Street, Newington, CT 06111, or call (800) 594-0200 or (203) 666-1541 [fax: (800) 594-0259].

A 5-W ATV system can transmit up to a distance of 30 to 40 miles, depending upon local radio interference, terrain, weather, etc.

## More models

With the experience gained in building this T-bot system, the reader can build other models. The company that makes the Erector Sets has revitalized and updated itself and brought a number of interesting kits to the market. The kits are called Meccano-Erector Sets that include motors, gears, and pulleys along with the standard Erector Set materials.

There are standard kits for building trucks, cars, motorcycles, land movers, etc. The kits provide a good springboard for building exotic T-bot explorers. Kits are available locally through Toys R Us dealerships.

## Parts list for the telepresence robot

- ☐ (1) Miniature color camera with 2.4-GHz transmitter and receiver—$99.95
- ☐ (1) Optional battery pack—$19.95
- ☐ (1) Radio-control system proportional control [two-channel receiver, two-channel transmitter, Xtals, 2 servo motors (42-oz torque)]—$62.95
- ☐ (1) 1-ft length of Velcro material—$4.50

Parts are available from:

Images SI Inc.
39 Seneca Loop
Staten Island, NY 10314
(718) 698-8305
www.imagesco.com

# 10

# Mobile platforms

PLATFORMS ARE THE FOUNDATION FOR MOBILE ROBOTS. There are two options: build or buy. If one has good mechanical ability or is willing to learn, building a platform from scratch offers distinct advantages. The platform is designed and built for the specific task and purpose of a robot. One has an almost unlimited choice of drive motors, gearboxes, mechanical linkage, power supplies, etc.

Buying a mobile platform relieves one from building a platform. But one is left with the gear ratio and power and speed designed for a different purpose. Here is a case in point: Most electric cars move too fast. If one doesn't have mechanical ability, this is the way to go. Typically one buys a radio-controlled (R/C) electric car. The radio controls are stripped from the unit. The electrical connections (wires) to the steering control and drive motor are retained.

Here are some things to keep in mind when purchasing an electric car for conversion. First, don't choose a car that's too small or lies too close to the ground. A small size will make it difficult to fit sensor systems and microcontrollers onto the chassis. If the car lies too close to the ground, it will get stuck easily. Choose a car with a high wheelbase.

Figure 10.1 shows an electric car that is a bad choice for turning into a mobile robot. It's too small to carry substantial weight, and notice how low to the ground it lies. This car will get stuck easily. Figure 10.2 shows a better choice. The platform is larger (can fit more components) and has a high wheelbase.

■ **10.1** *Small electric RC car unsuitable for conversion*

■ **10.2** *Large electric RC car suitable for conversion*

## Stepper motors

If one wants to build a platform, stepper motors make excellent drive motors. Some of the advantages of a stepper motor are as follows. Because a stepper motor turns in precise increments per step, a microcontroller can calculate the distance traveled by counting the clock pulses given to the stepper motor and knowing the diameter of the drive wheel. If two stepper motors are used on a mobile platform, one on each side, for locomotion and steering, precision turns are also possible. Because stepper motors are so important in robotics, we will look at the fundamental operation of stepper motors before we construct any circuits.

## Chapter ten

## Stepper motor construction and operation

Stepper motors are constructed using strong permanent magnets and electromagnets. The permanent magnets are located on the rotating shaft, called the *rotor.* The electromagnets or windings are located on the stationary portion of the motor, called the *stator.* Figure 10.3 illustrates a stepper motor stepping through one complete rotation. The stator, or stationary portion of the motor, surrounds the rotor.

In Fig. 10.3, position 1, we start with the rotor facing the upper electromagnet that is turned on. To move in a clockwise (CW) rotation, the upper electromagnet is switched off as the electromagnet to the right is switched on. This causes the rotor to rotate 90 degrees to align itself to the electromagnet in a CW rotation, shown in position 2. Continuing in the same manner, the rotor is stepped through a full rotation until we end up in the same position as we started, shown in position 5.

## Resolution

The degree of rotation per pulse is the resolution of the stepper motor. In the illustrated example of Fig. 10.3, the rotor turned 90 degrees per pulse, not a very practical motor. A practical stepper motor has a greater resolution (smaller steps), for instance, one that rotates its shaft 1 degree per pulse (or step). This motor requires 360 pulses (or steps) to complete one revolution. When a stepper motor is used for locomotion or positioning in a linear motion table, each step of the motor translates to a precise increment of linear movement.

Assume that one revolution of the motor is equal to 1" of linear travel. For a stepper motor that rotates 3.75 degrees per step, the increment of linear movement is approximately 0.01" per step. A stepper motor that rotates 1.0 degrees per step would give approximately 0.0027" per step. The increment of movement is inversely proportional to the degrees per step.

## Half stepping

It is possible to double the resolution of some stepper motors by a process known as *half stepping.* The process is illustrated in Fig. 10.4. In position I, the motor starts with the upper electromagnet switched on, as before. In position II the electromagnet to the right is switched on while keeping power to the upper coil on. Since both coils are on, the rotor is equally attracted to both electromagnets and positions itself in between both positions (a half

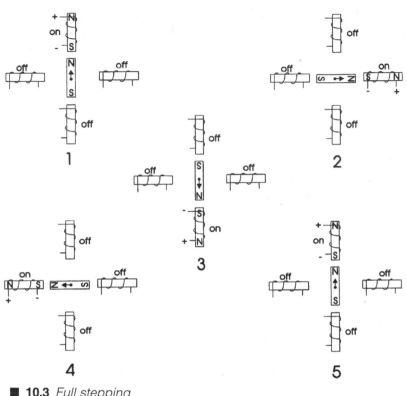

■ **10.3** *Full stepping*

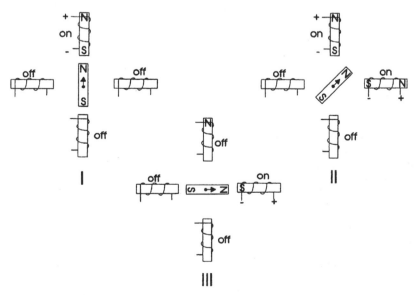

■ **10.4** *Half stepping*

step). In position III the upper electromagnet is switched off and the rotor completes one step. Although I am only showing one half step, the motor can be half stepped through the entire rotation.

## Other types of stepper motors

There are four-wire stepper motors. These stepper motors are called *bipolar* and have two coils, with a pair of leads to each coil. Although the circuitry of this stepper motor is simpler than the one we are using, it requires a more complex driving circuit. The circuit must be able to reverse the current flow in the coils after it steps.

## Real world

The stepper motor illustrated rotated 90 degrees per step. Real-world stepper motors employ a series of mini-poles on the stator and rotor. The mini-poles reduce the degrees per step and improve the resolution of the stepper motor. Although the drawing in Fig. 10.5 appears more complex, its operation is identical to the previous illustrations shown in Figs. 10.3 and 10.4.

The rotor in Fig. 10.5 is turning in a CW rotation. In the first position the north pole of the permanent magnet on the rotor is aligned with the south pole of the electromagnet on the stator. Notice that there are multiple positions that are all lined up. In the second position the electromagnet is switched off and the coil to its immediate left is switched on. This causes the rotor to rotate CW by a precise amount. It continues in this same manner for all the steps. After eight steps, the sequence of electric pulses would start to repeat. Half stepping with the multipole position is identical to the half step described before.

Figure 10.6 is an electric equivalent circuit of a unipolar stepper motor. The stepper motor has six wires coming out from the casing.

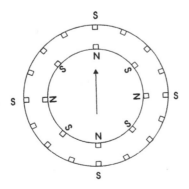
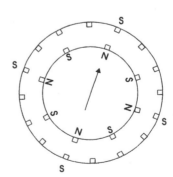

■ **10.5** *Multipole operation*

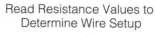

Read Resistance Values to
Determine Wire Setup

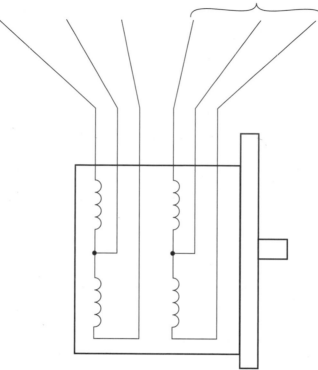

■ **10.6** *Schematic of six-wire unipolar stepper motor*

We can see from Fig. 10.6 that three leads go to each half of the coil windings and that the coil windings are connected in pairs. If you just picked this stepper motor and didn't know anything about it, the simplest way to analyze it would be to check the electrical resistance between the leads. By making a table of the wire colors and resistances measured between the leads you would quickly find which wires were connected to which coils. (In some cases a unipolar stepper motor will only have five wires coming out of it. In this case the center taps of the coils are wired together.)

The motor we are using has a 110-ohm resistance between the center tap wire and each end lead and a 220-ohm resistance between the two end leads. A wire from each of the separate coils will show an infinitely high resistance (no connection) between them. Armed with this information you can just about tackle any six-wire stepper motor you come across. The stepper motor we are using rotates 1.8 degrees per step.

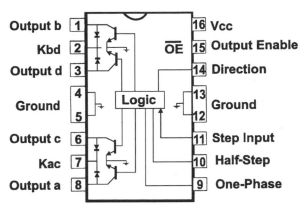

■ **10.7** *UCN-5804 stepper motor controller chip*

## UCN-5804

Figure 10.7 is a schematic pin-out of the UCN-5804. This integrated circuit (IC) is designed to control and drive a four-phase unipolar stepper motor, such as the one we are using. Features of the UCN-5804 are as follows:

☐ 1.25-ampere (A) maximum output current (continuous)

☐ 35-volt (35V) output sustaining voltage

☐ Full-step and half-step outputs

☐ Output enable and direction control

☐ Internal clamp diodes

☐ Power-on reset

☐ Internal thermal shutdown circuitry

The IC has a continuous output rating of 1.25 A per phase at a maximum voltage of 35 V. This is more than enough power to run our 12V stepper motor. The current required per phase (12 V/110 ohms = 0.11 A) is about one-tenth of an ampere.

The UCN-5804 internal logic sequences its output pins in time with a square wave pulse delivered to pin 11. Each square wave pulse (high to low transition) delivered to this pin increments the stepper motor sequence.

When you reach the end of your table, the sequence repeats starting from the top of the table. To reverse the stepper motor direction, start the sequence from the bottom and work toward the top.

Pin 15 is the output enable. When this pin is held high, all outputs on the IC are disabled (off). If this function isn't required by your circuit or system, this pin should be tied to ground (low).

■ **Table 10.1 Full-Step Sequence**

| a | b | c | d | |
|---|---|---|---|---|
| On | — | — | — | Output pins of UCN-5804 |
| — | On | — | — | (see Fig. 10.7) |
| — | — | On | — | |
| — | — | — | On | |

■ **Table 10.2 Half-Step Sequence**

| a | b | c | d | |
|---|---|---|---|---|
| On | — | — | — | Output pins of UCN-5804 |
| On | On | — | — | |
| — | On | — | — | |
| — | On | On | — | |
| — | — | On | — | |
| — | — | On | On | |
| — | — | — | On | |
| On | — | — | On | |

Pin 14 is the direction. When this pin is tied low or connected to ground, it will follow the sequence in either Table 10.1 or 10.2 starting from the top line and working downward. When this pin is tied high (15 V), it will reverse the sequence direction starting from the bottom and working its way to the top.

## Using the UCN-5804

Figure 10.8 is a schematic using the UCN-5804. The clocking signal is provided by the 555 timer. The clocking signal may be increased or decreased using potentiometer V1. Varying the frequency of the clock signal directly controls the speed of the stepper motor. In this chapter we show how the PIC microcontroller can drive a stepper motor with or without specialty components.

In this schematic, three manual on/off switches control additional functions. These pins that the switches are connected to can also be controlled by the input/output (I/O) pins off the basic stamp microcontroller. The switch connected to pin 15 is the enable pin. When brought high, this pin disables the output of the UCN-5804 chip, stopping the stepper motor.

The switch connected to pin 14 controls the shaft's direction, CW or counterclockwise (CCW). A switch connected to pin 10 controls the step/half-step function of the UCN-5804. When pin 10 is

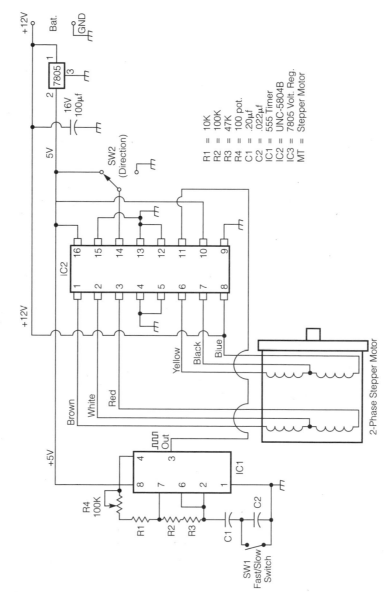

■ **10.8** *Schematic of basic stepper motor driver circuit*

**221**

brought high, the chip operates in the half-step mode. This mode doubles the resolution of the stepper motor. For instance, the motor we are using rotates the shaft 1.8 degrees per step. When operating in the half-step mode, the shaft rotates 0.9 degrees per step and the overall rotation speed [revolutions per minute (rpm)] of the shaft will be one-half of the speed of the full-step mode. If pin 10 is brought to ground, the UCN-5804 will operate in full-step mode.

## Connecting a wheel to a stepper motor shaft

Connecting a drive wheel to a shaft can become a major problem. A simple solution is provided (see Fig. 10.9). Purchase a large-diameter plastic gear with a set screw. The mounting hole on the gear should match the shaft diameter of the stepper motor. Center the gear on the wheel. Drill three holes, 120 degrees apart, through the gear and wheel. Mount the wheel to the gear using three machine screws, washers, and nuts. Next mount the wheel gear assembly to the stepper motor shaft using the set screw.

## Building a stepper microcontroller

Now let's build a simple stepper motor controller from a PIC16F84 and examine the operating principles of stepper motors.

### First stepper circuit

Figure 10.10 is the schematic for our first test circuit. The output lines from the PIC16F84 are buffered using a 4050 hexadecimal (hex) buffer chip. Each buffered signal line is connected to an NPN

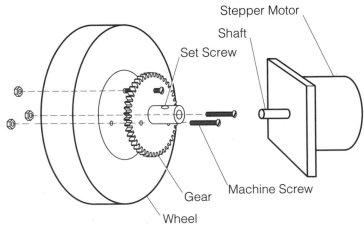

■ **10.9** *Connecting wheel to motor shaft*

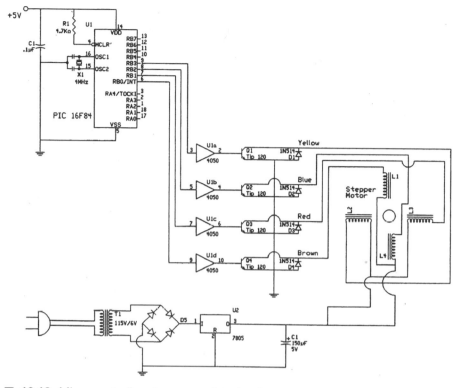

■ **10.10** *Microcontroller stepper motor circuit*

transistor. The TIP 120 transistor is actually an NPN Darlington; in the schematic it is shown as a standard NPN. TIP 120 transistors act like switches, turning on one stepper motor coil at a time.

The diodes placed across each transistor protect the transistor from the inductive surge created when switching current on and off in the stepper motor coils. The diode provides a safe return path for the reverse current. Without the diodes, the transistor will be more prone to failure and/or a shorter life.

## Stepper motors

Figure 10.11 is an electric equivalent circuit of the stepper motor we are using. The stepper motor has six wires coming out from the casing.

Let's assume you just picked this stepper motor and didn't know anything about it. As stated before, the simplest way to analyze the motor is to check the electrical resistance between the leads. By making a table of the resistances measured between the leads you'll quickly find which wires are connected to which coils.

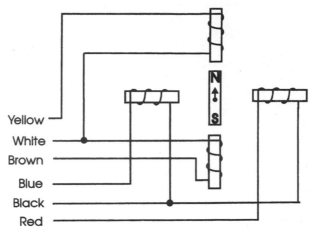

**■ 10.11** *Wiring of unipolar stepper motor*

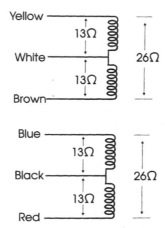

**■ 10.12** *Resistance reading from leads of unipolar stepper motor*

Figure 10.12 shows how the resistance of the motor we are using looks. There is a 13-ohm resistance between the center tap wire and each end lead, and 26 ohms between the two end leads. The resistance reading from wires originating from separate coils will show an infinitely high resistance (no connection). For instance, this would be the case when reading the resistance between the blue and brown leads. Armed with this information, you can wire it properly into a circuit.

### First test circuit and program

After you are finished constructing the test circuit, program the PIC with the following BASIC program. The program is kept small

## ■ Table 10.3 Full-Step Transistors

| Q1 | Q2 | Q3 | Q4 | Port B Output (decimal) |
|----|----|----|----|-------------------------|
| On | — | — | — | 1 |
| — | On | — | — | 2 |
| — | — | On | — | 4 |
| — | — | — | On | 8 |

and simple to show how easy it is to get a stepper motor moving. Table 10.3 shows that each step in the sequence turns on one transistor. Use Table 10.3 to follow the logic in the PICBASIC program. When you reach the end of the table, the sequence repeats starting back at the top of the table.

```
'Stepper Motor Controller
Symbol TRISB = 134    'Initialize TRISB to 134
Symbol PortB = 6    'Initialize portb to 6
symbol ti = b6    'Initial ti delay
ti = 25    'Set delay to 25 ms
Poke TRISB,0    'Set PORTB lines output
start:    'Forward rotation sequence
poke portb,1    'Step 1
pause ti    'Delay
poke portb,2    'Step 2
pause ti    'Delay
poke portb,4    'Step 3
pause ti    'Delay
poke portb,8    'Step 4
pause ti    'Delay
goto start    'Do again
```

### One rotation

Using whole steps, the stepper motor requires 200 pulses to complete a single rotation (360 degrees/1.8 degrees per step). Having the PIC microcontroller count pulses allows it to control and position the stepper motor's rotor.

## Second PICBASIC program

This second PICBASIC program is far more versatile. The user can modify programmed parameters (time delay) as the program is running using one of the four switches connected to port A. Pressing switch 1 (SW1) lengthens the delay pause between steps in the sequence and subsequently makes the stepper motor rotate slower. Pressing SW2 has the opposite effect. If you press SW3, the program halts the stepper motor and stays in a holding loop for as long as SW3 is closed (or pressed). Rotation direction, CW or CCW, is controlled with SW4. Pressing SW4 reverses the stepper motor direction. The direction stays in reverse for as long as SW4 is pressed (or closed).

```
'Stepper motor controller
Symbol TRISB = 134    'Initialize TRISB to 134
Symbol TRISA = 133    'Initialize TRISA to 133
Symbol PortB = 6    'Initialize portb to 6
Symbol PortA = 5    'Initialize porta to 5
symbol ti = b6    'Initial ti delay
ti = 100    'Set delay to 100 ms
Poke TRISB,0    'Set PORTB lines output
start:    'Forward stepper motor rotation sequence
poke portb,1    'Step 1
pause ti    'Delay
poke portb,2    'Step 2
pause ti    'Delay
poke portb,4    'Step 3
pause ti    'Delay
poke portb,8    'Step 4
pause ti    'Delay
goto check    'Jump to check switch status
start2:    'Reverse motor rotation sequence
poke portb,8    'Step 1
pause ti    'Delay
poke portb,4    'Step 2
pause ti    'Delay
poke portb,2    'Step 3
pause ti    'Delay
poke portb,1    'Step 4
pause ti    'Delay
goto check    'Jump to check switch status
Check:    'Switch status
Peek PortA, B0    'Peek the switches
If bit0 = 0 then loop1    'If SW1 is closed, increase ti
if bit1 = 0 then loop2    'If SW2 is closed, decrease ti
if bit2 = 0 then hold3    'Stop motor
if bit3 = 0 then start    'Go forward
goto start2    'Go reverse
loop1:    'Increase delay
poke portb,0    'Turn off transistors
ti = ti + 5    'Increase delay by 5 ms
pause 50    'Delay
```

```
if ti > 250 then hold1    'Limit delay to 250 ms
peek porta,b0   'Check switch status
if bit0 = 0 then loop1    'Still increasing delay?
goto check    'If not, jump to main switch status
loop2:   'Decrease delay
poke portb,0   'Turn off transistors
ti = ti - 5    'Decrease delay by 5 ms
pause 50    'Pause a moment
if ti < 20 then hold2    'Limit delay to 20 ms
peek porta,b0    'Check switch status
if bit1 = 0 then loop2    'Still decreasing delay?
goto check    'If not, jump to main switch status
hold1:    'Limit upper delay
ti = 245    'to 250 ms
goto loop1   'Go back
hold2:   'Limit lower delay
ti = 25    'to 25 ms
goto loop2   'Go back
hold3:   'Stop stepper motor
poke portb,0   'Turn off transistor
peek porta, b0    'Check switches
if bit2 = 0 then hold3    'Keep motor off?
goto check    'If not, go to main switch status
check
```

The schematic for this program is shown in Fig. 10.13. In the photograph of the circuit (see Fig. 10.14), the four switches are difficult to make out. There are the four bare wire strips behind the PIC microcontroller. The top sides of the bare wire strips are connected to +5V through 10K-ohm resistors. A wire from each switch is connected to the appropriate pin on port A. A single wire is connected to ground and is used to close any of the switches by touching the bare wire strip.

### Half stepping

Half stepping the motor will effectively double the resolution. In this instance it requires 400 pulses to complete one rotation. Table 10.4 shows the switching logic needed in a program. When you reach the end of the table, the sequence repeats starting back at the top of the table.

### The ti delay variable

The ti variable used in each PICBASIC program controls a delay pause whose purpose is to slow down the output sequence to port B. Without the pause, the sequence may run too fast for the stepper motor to respond, causing the stepper motor to malfunction.

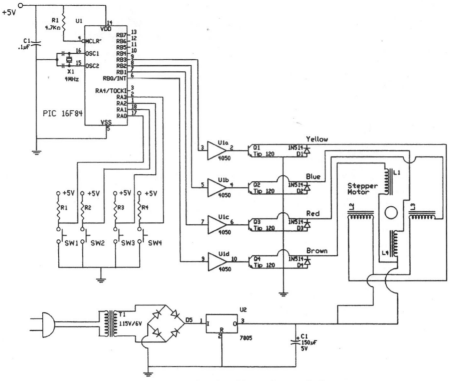

■ **10.13** *Microcontroller stepper circuit with option switches*

You may want to vary the `ti` variable in the program depending upon your PIC crystal speed. You can experiment with the `ti` variable until you find the best range for your particular PIC.

## Troubleshooting

If the motor doesn't move at all, check the diodes. Make sure you have them in properly, facing in the direction shown in the schematic.

If the stepper motor moves slightly and/or quivers back and forth, there are a number of possible causes.

1. If you are using a battery power supply, the batteries may be too weak to power the motor properly. *Note:* Batteries wear out quickly because the current draw from stepper motors is usually high.

2. If you substituted the TIP 120 NPN transistor for another transistor, the substitute transistor may not be switching

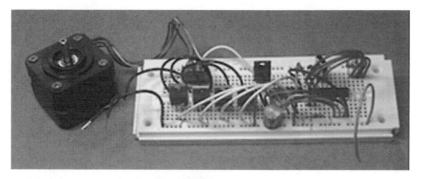

■ **10.14** *Microcontroller stepper motor circuit*

■ **Table 10.4 Half-Step Transistor**

| Q1 | Q2 | Q3 | Q4 | Port B Output (decimal) |
|----|----|----|----|-------------------------|
| On | — | — | — | 1 |
| On | On | — | — | 3 |
| — | On | — | — | 2 |
| — | On | On | — | 6 |
| — | — | On | — | 4 |
| — | — | On | On | 12 |
| — | — | — | On | 8 |
| On | — | — | On | 9 |

properly or the current load of the stepper motor may be too great. *Solution:* Use TIP 120 transistors.

3. You have the stepper motor improperly wired into the circuit. Check the coils using an ohmmeter and rewire if necessary.

4. The pulse frequency is too high. If the pulses to the stepper motor are going faster than the motor can react, the motor will malfunction. The pulse frequency is controlled by the $t i$ variable in the program. Increasing the value of this variable will slow down the pulse frequency to the stepper motor. The solution to this is to reduce the pulse frequency.

## Using a PIC microcontroller and a UCN-5804 stepper motor IC

We have controlled the stepper motor directly from the PIC chip. We have also built a stepper motor controller using dedicated stepper ICs. By incorporating stepper motor controller chips into the PIC microcontroller design, we can derive benefits from both

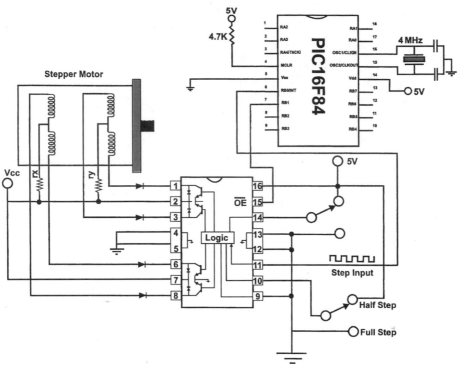

**■ 10.15** *Schematic of microcontroller and stepper motor IC*

components. The UCN-5804 does most of the grunt work of controlling a stepper motor. This simplifies our microcontroller program and overall circuit while enhancing the hardware, a good combination.

The schematic for a stepper motor controller using a dedicated IC is shown in Fig. 10.15, and a photograph of the circuit is shown in Fig. 10.16. The UCN-5804 is powered by a 5V direct current (DC) power supply. While internally powered by 5 V, it can control stepper motor voltages up to 35 V.

Notice in the schematic that there are two resistors labeled "rx" and "ry" that do not show any resistance value. Depending upon the stepper motor, these resistors may or may not be necessary. Their purpose is to limit current through the stepper motor to 1.25 A (if necessary).

Let's look at our 5V stepper motor. It has a coil resistance of 13 ohms. The current draw of this motor will be 5 V/13 ohms = 0.385 A, or 382 milliamperes (mA), well below the 1.25-A maximum rating of the UCN-5804. So in this case resistors rx and ry are not needed and may be eliminated from the schematic.

Before we move on, let's look at one more case. A 12V stepper motor has a phase (coil) resistance of 6 ohms. The current drawn by this motor is 12 V/6 ohms = 2 A. This is above the UCN-5804 maximum current rating. To use this stepper motor, you must add the rx and ry resistors. The rx and ry resistor values should be equal to each other, so each phase will have the same torque. The values chosen for these resistors should limit the current drawn to 1.25 A or less. In this case the resistors should be at least 4 ohms [5 to 10 watts (W)]. With the resistors in place the current drawn is 12 V/10 ohms = 1.20 A.

The inputs to the UCN-5804 are compatible with complementary metal-oxide semiconductor (CMOS) and transistor-transistor logic (TTL). This means we can connect the outputs from our PIC microcontroller directly to the UCN-5804 and expect it to function properly. The step input (pin 11) to the UCN-5804 is generated by the PIC microcontroller. The output enable pin when held low enables the stepper motor; when brought high, it disables (stops) the stepper motor.

Pins 10 and 14 on the UCN-5804 are controlled by switches that bring the pins to a logic high or low. Pin 10 controls whether the output to the stepper motor will be full-step or half-step, and pin 14 controls direction. If we want, these options may also be put under the PIC control. The pins are brought to a logic high or low to activate the options just like the output enable pin.

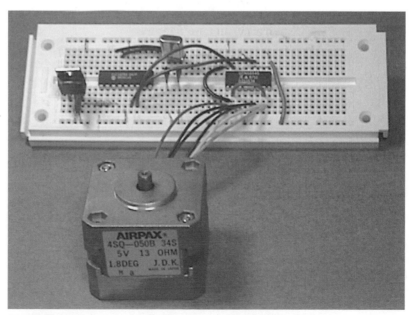

■ **10.16** *Microcontroller and stepper motor IC*

The following is a PICBASIC program that uses a dedicated stepper motor IC.

```
'Stepper motor with UCN-5804
Symbol TRISB = 134    'Initialize TRISB to 134
Symbol PortB = 6    'Initialize portb to 6
Poke TRISB,0    'Set PORTB lines output
low1    'Bring output enable low to run
start:
pulsout 0, 10000    'Send 10-ms pulse to UCN-5804
goto start    'Do again
```

In this case I again wrote a simple core program to show how easy it is to get the stepper motor running. You can, of course, add options to the program to change the pulse frequency, connect the direction and step mode pins, etc.

## Parts list for the stepper motor controller

- ☐ (1) 16F84 microcontroller
- ☐ (2) 22-picofarad (pF) capacitors
- ☐ (1) 4.0-MHz crystal
- ☐ (1) 4.7K-ohm, $1/4$-W resistor
- ☐ (1) 555 timer
- ☐ (1) UCN-5804B stepper motor controller chip
- ☐ (1) Stepper motor [unipolar (six-wire)]
- ☐ (1) Step-down wall transformer
- ☐ (6) 1N914 diodes
- ☐ (4) TIP 120 NPN transistors
- ☐ (1) Voltage regulator (7805, 7812)
- ☐ (1) Rectifier 50V, 1 A
- ☐ (1) 150-μF capacitor
- ☐ (1) 4050 hex buffer chip
- ☐ Miscellaneous: Solderless breadboard

Parts are available from: Images Company, James Electronics, JDR MicroDevices, and Radio Shack. See Suppliers at the end of the book.

# 11

# Walker robots

WALKERS ARE A CLASS OF ROBOTS THAT IMITATE THE locomotion of animals and insects. Essentially, walker robots use legs for locomotion. Locomotion by legs is hundreds of millions of years old. In contrast to this, wheels are relatively a new science, being only 7000 to 10,000 years old. Wheels are good, but they require a relatively smooth surface to ride upon. Just look at an aerial photograph of any city or suburb to see the highways and streets crisscrossing the landscape.

## Why build walkers?

Walker robots have the potential to transverse rough terrain that is impassable by standard wheeled vehicles. It is with this in mind that robotists are developing walker robots.

## Imitation of life

Sophisticated walkers imitate insects, crabs, and sometimes humans. Bipedal walkers are rare, requiring a good deal of engineering science. I plan to have a bipedal walker robot project in my next book on robotics, tentatively titled *Pic-Robotics.* In this chapter we will build a six-legged walker robot.

## Six legs—tripod gait

Using a six-legged model we can demonstrate the famous tripod gait used by the majority of legged creatures. In the following drawings a dark circle means the foot is firmly planted on the ground and supporting the weight of the creature. A light circle means the foot is up and movable.

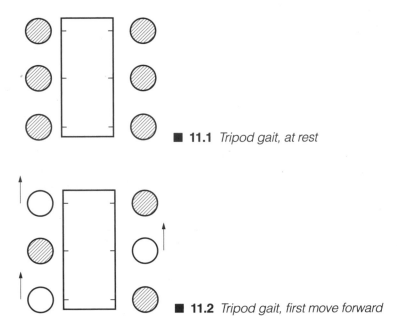

■ **11.1** *Tripod gait, at rest*

■ **11.2** *Tripod gait, first move forward*

Figure 11.1 shows our creature at rest. All feet are on the ground. From the resting position our creature decides to move forward. To step forward, it lifts three of its legs (see Fig. 11.2, white circles), leaving its weight on the remaining three legs (dark circles). Notice that the legs supporting the weight (dark circles) are in the shape of a tripod. This is a stable weight-supporting position. Our creature is unlikely to fall over. The three lifted legs (white circles) are free to move, and they move forward.

Figure 11.3 illustrates where the three lifted legs move. At this point, the creature's weight shifts from the stationary legs to the movable legs (see Fig. 11.4). Notice that the creature's weight is still supported by a tripod position of legs. Now the other set of legs move forward and the cycle repeats. This is called a *tripod gait*, because the creature's weight is always supported by a tripod positioning of legs.

## Creating a walker robot

There are a lot of little wind-up toy walkers around. These toy walkers move their legs up and down, back and forth, using a rotary cam mechanism. While these walkers work, and some are surprisingly fast, our task is to build a walker that does not use a rotary cam to imitate walking.

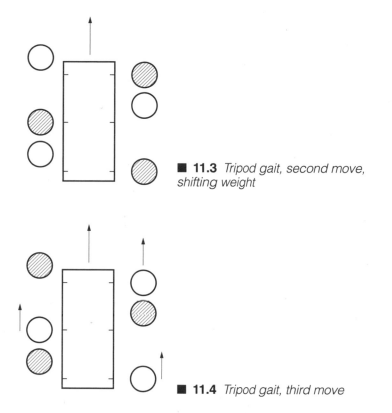

■ **11.3** *Tripod gait, second move, shifting weight*

■ **11.4** *Tripod gait, third move*

We will build a walker robot that imitates a tripod gait. The walker outlined in this chapter requires a minimum of three servo motors. There are numerous hexapod and quadrapod walker designs that require greater freedom of movement per leg. Greater freedom of movement per leg means more independent drivers per leg. If one is using servo motor drivers, this can be achieved using two, three, or four servo motors per leg.

The need for so many servo motors (drivers) is because each leg on the walker needs to have a minimum of two axes (degrees) of freedom. One to move up or down and the second to move (swing) forward and back.

### Three-servo walker

The walker robot we will make is a compromise in design and construction, but requires only three servo motors. Even so, using just three servo motors, it is a true tripod gait walker. Our walker uses three lightweight HS300 servo motors [42-ounce (oz) torque] and a 16F84-04 microcontroller.

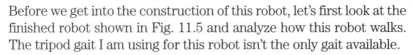

■ **11.5** *Hexapod walker ready to run*

## Function

Before we get into the construction of this robot, let's first look at the finished robot shown in Fig. 11.5 and analyze how this robot walks. The tripod gait I am using for this robot isn't the only gait available.

At the front of the robot we have two servo motors. Each servo motor controls the two legs on its side, the front leg and the back leg. The front leg is attached directly to the rotor of the servo motor. It is capable of swinging forward and backward. The back leg connects to the front leg through a linkage. The linkage makes the back leg follow the action of the front leg as it moves forward and back.

The two center legs are controlled by a third servo motor. This servo motor rotates the center legs 20 to 30 degrees in a clockwise (CW) or counterclockwise (CCW) rotation that tilts the robot to the left or right.

With this information under our belt we can now look to see how our robot will walk. Look at Fig. 11.6. We start in the rest position. Each circle represents a leg. As before, the dark circles show the weight-bearing legs. Notice in the rest position, the center legs do not support any weight. These legs are $1/8$" shorter than the front and back legs.

In position A the center legs are rotated CW by about 20 degrees from center position. This causes the robot to tilt to the right. The weight distribution is now on the front and back right legs and the center left leg. This is the standard "tripod" position described earlier. Since there is no weight on the front and back left legs, they are free to move forward as shown in the B position of Fig. 11.6.

In the C position the center legs are rotated CCW by about 20 degrees from center position. This causes the robot to tilt to the left. The weight distribution is now on the front and back left legs and the center right leg. Since there is no weight on the front and back right legs, they are free to move forward as shown in the D position.

In the E position the center legs are rotated back to their center position. The robot is not in a tilted position, so its weight is distributed on the front and back legs. In the F position, the front and back legs are moved backward simultaneously, causing the robot to move forward. The walking cycle then repeats.

This is the first gait pattern I tried, and it worked. There are other walking patterns you can design, develop, and experiment with. I will leave it to you to develop walking patterns for reverse (walking backward), turning left, and turning right. In my next book on robotics, I will continue the development of this robot, providing wall and

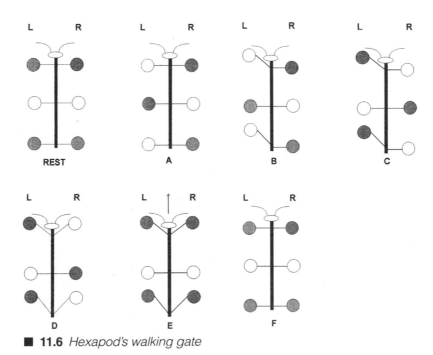

■ **11.6** *Hexapod's walking gate*

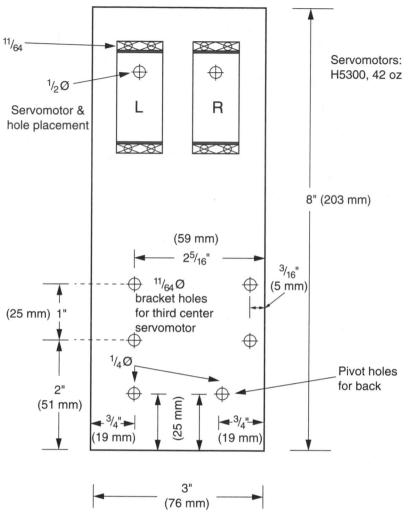

**■ 11.7** *Main body*

collision sensors as well as providing the ability to walk backwards and turn.

## Construction

For the main body I used a sheet of aluminum 3" wide, 8" long, and 0.032" thick. The servo motors are mounted to the front of the body (see Fig. 11.7). The drawings of the servo motor holes shown should be photocopied and taped to the aluminum sheet. The photocopy will provide accurate hole locations for mounting the servo motors.

The four $^{11}/_{64}$"-diameter holes a little past halfway down the main body are for mounting the center servo motor. These four holes are

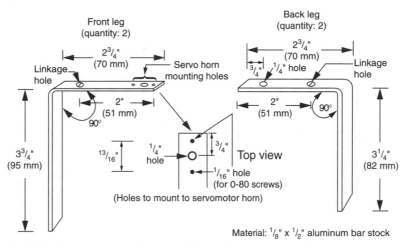

Front leg
(quantity: 2)

Back leg
(quantity: 2)

Linkage hole

Servo horn
mounting holes

Linkage hole

$2^3/_4$"
(70 mm)

$2^3/_4$"
(70 mm)

$3/_4$"  $1/_4$" hole

2"
(51 mm)

2"
(51 mm)

90°

90°

$3^3/_4$"
(95 mm)

$3^1/_4$"
(82 mm)

$^{13}/_{16}$"

$1/_4$"
hole

$3/_4$"

Top view

$^1/_{16}$" hole
(for 0-80 screws)

(Holes to mount to servomotor horn)

Material: $1/_8$" x $1/_2$" aluminum bar stock

■ **11.8** *Construction of front and back legs*

offset to the right side. This is necessary to align the servo motor's horn in the center of the body. The bottom two holes are for mounting the pivots for the two back legs.

Use a punch to dimple the metal in the center of each hole you plan to drill. This will prevent the drill bit from walking when you drill the hole. If you don't have a punch available, use the pointed tip of a nail for a quick substitute.

The legs for the robot are made from $1/_2$" wide $\times$ $1/_8$" thick aluminum bar stock (see Fig. 11.8). There are four holes that are drilled into the two front legs. The back legs only need two holes each, one for the pivot and the other for the linkage. Also notice that the back legs are $1/_4$" shorter than the front legs. This compensates for the height of the servo motor mounting horn on the front servo motors where the front legs are attached. Shortening the back legs makes the robot platform level.

After the holes are drilled, we need to bend the aluminum bar into shape. Secure the aluminum bar in a vise $2^3/_4$" from the end with the drilled holes. Apply pressure to bend the aluminum bar at a 90 degree angle. It's best to apply pressure at the base of the aluminum bar close to the vise. This will bend the leg at a 90 degree angle, while keeping the lower portion of the leg straight without any bowing of the lower portion.

The center legs are made from one piece of aluminum (see Fig. 11.9). The center legs are about $1/_8$" shorter than the front and back legs when mounted to the robot. So when centered, the legs do not support any weight. These legs are for tilting the robot to

*239*

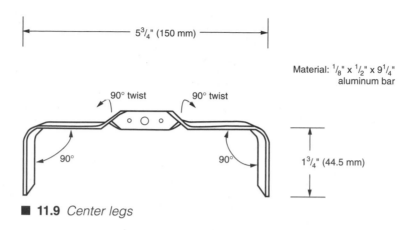

■ **11.9** *Center legs*

the left or right. The legs tilt the robot by rotating the center servo motor approximately ±20 degrees.

To produce the center legs, first drill the mounting holes in the center of the $\frac{1}{8}$" × $\frac{1}{2}$" × $9\frac{1}{4}$" aluminum bar. Secure the aluminum bar in a vise. The top of the vise should hold the aluminum bar $\frac{3}{4}$" from the center of the aluminum bar. Grab the aluminum bar with pliers about $\frac{1}{2}$" above the vise. Keeping a secure grip with the pliers, slowly twist the aluminum bar 90 degrees. Don't go fast or you could easily snap the aluminum bar. Repeat the twist on the other side.

After the two 90 degree twists have been made, make the other 90 degree bend for the legs as we have done before for the front and back legs.

## Mounting the servo motors

The front servo motors are attached to the aluminum body using plastic 6-32 machine screws and nuts. The reason I am using plastic screws is that they are a little flexible, allowing the drilled holes to be slightly off center from the mounting holes on the servo motor.

The legs are attached to the servo motor's plastic horn. For this I used 0-80 machine screws and nuts. When mounting the servo motor horn on the servo motor, make sure that each leg can swing forward and backward an equal amount from a perpendicular position.

## Linkage

The linkage between the front and back legs is made from 4-40 threaded rod (see Fig. 11.10). In the prototype robot the linkage is $5\frac{3}{4}$" center to center. The linkage fits inside the holes in the

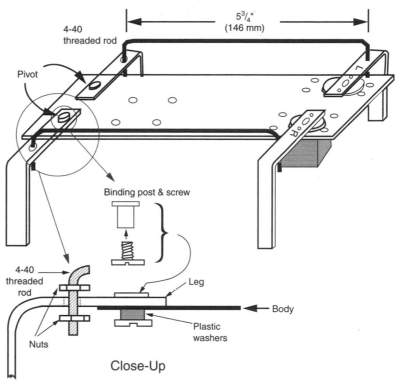

**5³/₄"**
**(146 mm)**

4-40
threaded rod

Pivot

Binding post & screw

4-40
threaded
rod

Leg

Body

Plastic
washers

Nuts

**Close-Up**

■ **11.10** *Close-up of pivot and linkage*

front and back legs. The linkage may be secured using a few 4-40 hex nuts.

The back legs must be attached to the body of the robot before you make the linkage. The pivot for the back legs is made from a ³/₈" binding post and screw. The leg is attached as shown in the close-up in Fig. 11.10. The plastic washers underneath the body are necessary. They fill up the space between the aluminum body and the bottom of the screw. This keeps the leg close to the aluminum body without sagging. I chose plastic washers for less friction. Do not use so many washers that force is created binding the leg to the body. The joint should pivot freely. Look at Figs. 11.11 and 11.12 for pictures of our hexapod walker robot thus far.

## Center servo motor

Attaching the center servo motor to the body requires two L-shaped brackets (see Fig. 11.13). Drill the holes in the aluminum stock, and then bend at a 90 degree angle to form the L brackets. Attach the two L brackets to the center servo motor using the plastic

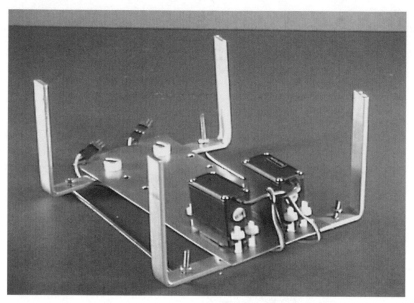

■ **11.11** *Underside of hexapod with two front servo motors*

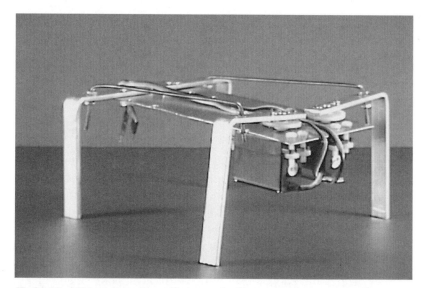

■ **11.12** *Midconstruction of hexapod with two front servo motors*

screws and nuts (see Fig. 11.14). Next mount the center servo motor assembly under the robot body. Align the four holes in the body with the top holes in the L brackets. Secure with plastic screws and nuts. Figures 11.15 and 11.16 show the under side and top side of the hexapod robot.

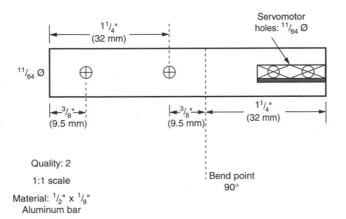

Servomotor
holes: $^{11}/_{64}$ Ø

$1^1/_4$"
(32 mm)

$^{11}/_{64}$ Ø

$^3/_8$"
(9.5 mm)

$^3/_8$"
(9.5 mm)

$1^1/_4$"
(32 mm)

Quality: 2

1:1 scale

Material: $^1/_2$" x $^1/_8$"
Aluminum bar

Bend point
90°

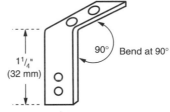

$1^1/_4$"
(32 mm)

90°    Bend at 90°

■ **11.13** *Center servo motor bracket*

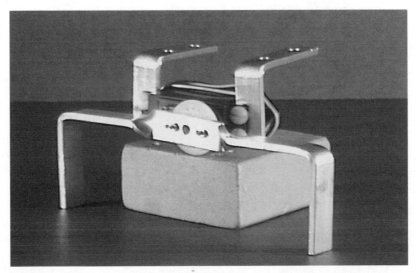

■ **11.14** *Center servo motor with brackets and center legs attached*

## Electronics

Figure 11.17 shows the schematic for the servo motors and PIC microcontroller. Notice the 6V battery pack is powering the micro-controller as well as the servo motors. The battery pack is 16V

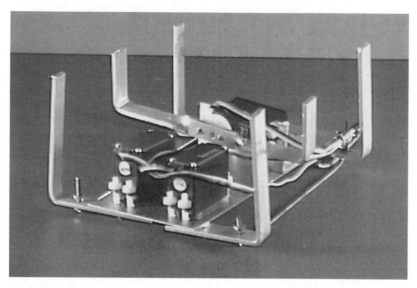

■ **11.15** *Underside of hexapod walker with three servo motors*

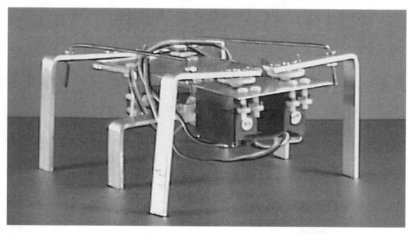

■ **11.16** *Hexapod walker ready for control electronics*

using 4 AA batteries. The microcontroller circuit is built on a small solderless breadboard. The battery pack and circuit are laid on top of the aluminum body. Figure 11.5 shows the completed walker ready to run.

## Microcontroller program

The 16F84 microcontroller controls the three servo motors. There are plenty of input/output (I/O) lines and programming space left over to improve and add to this basic walker.

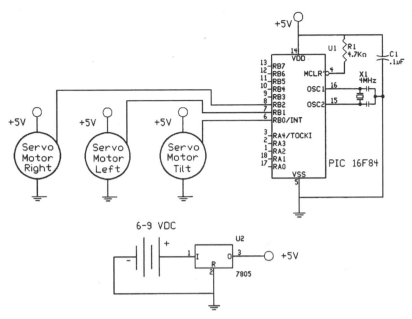

■ **11.17** *Schematic of hexapod walker*

## PICBASIC program

```
'Hexapod walker
' The connections
' Left servo motor      Pin RB1
' Right servo motor     Pin RB2
' Tilt servo motor      Pin RB0
' Moves in forward direction only
start:
FOR B0 = 1 TO 60
        pulsout 0,155    ' Start by tilting CW, lift right side
        pulsout 1,145    ' Keep left legs back
        pulsout 2,145    ' Bring right legs forward
        pause 18
NEXT B0
FOR B0 = 1 TO 60
        pulsout 0,190    ' Tilt CCW, lift left side
        pulsout 1,200    ' Bring left legs forward
        pulsout 2,145    ' Keep right legs forward
        pause 18
NEXT B0
for b0 = 1 to 15
        pulsout 0,172    ' No tilt
        pulsout 1,200    ' Keep left legs forward
```

```
        pulsout 2,145      ' Keep right legs forward
        pause 18
next b0
for B0 = 1 to 60
        pulsout 0,172      ' No tilt
        pulsout 1,145      ' Bring left legs back
        pulsout 2,200      ' Bring right legs back
        pause 18
next b0
        goto start
```

Not all servo motors are exactly alike or respond in an identical manner to the same `pulsout` command. The servo motors you purchase to build this robot will probably vary somewhat from the servo motors I used. Keeping this in mind, the `pulsout` commands that control the position of the servo motors may need to be adjusted. Adjust the numerical value of the `pulsout` commands to compensate for the particular servo motors used in your hexapod robot walker.

While this PICBASIC program only provides for forward motion, a little experimentation on the part of robotists can have this robot turning to the left or right and walking backward. A few sensor switches on the front can inform the robot when it has encountered an obstacle.

## Parts list for the walker robot

☐ Servo motors

☐ 16F84 microcontrollers

☐ Aluminum bars

☐ Aluminum sheets

☐ 4-40 threaded rods and nuts

☐ Plastic machine screws, nuts, and washers

Parts are available from:

Images Company
39 Seneca Loop
Staten Island, NY 10314
(718) 698-8305

http://www.imagesco.com

# 12

# Solar-ball robot

THE INSPIRATION FOR THIS ROBOT ORIGINALLY BEGAN WITH Richard Weait of North York, Toronto. Richard created a light-seeking robot in a transparent globe (ball). More recently, Dave Hrynkiw from Calgary, Canada, picked up the ball (so to speak) and developed a series of light-seeking mobile solar-ball robots.

There are two features to this mobile robot that are interesting (see Fig. 12.1). First is the method of locomotion. Inside the globe is a gearbox. One end of the gearbox's shaft is secured and locked to the inside of the inner surface of the transparent globe. The shaft being locked cannot rotate, which forces the gearbox itself to rotate. The gearbox is heavy, which moves the center of gravity of the sphere forward. In doing so, the sphere moves forward.

When at rest, the weight of the gearbox keeps it at bottom dead center (the gearbox facing down), and the ball resists rolling. When the gearbox is activated, the box begins to rotate inside the globe. This moves the center of gravity of the ball forward, causing the ball to roll forward.

The second feature relates to the power supply for the gearbox. The original solar robots had an onboard power supply that provided intermittent power to the gearbox. (For more information on this type of power supply, see Chap. 3.) The onboard power supply consists of a solar cell, a main capacitor, and a slow oscillating or trigger circuit. When exposed to sunlight, the solar cell begins charging the circuit's main capacitor. When the capacitor reaches a certain voltage, a trigger circuit dumps the stored electricity through a high-efficiency motor connected to the gearbox, causing the robot to move forward a little.

■ **12.1** *Solar-ball robot*

This solar-ball robot uses a similar gearbox assembly, but for power uses two standard AA batteries. The disadvantage to batteries is that they must be replaced when worn out. The advantage, however, is that they supply continuous power to the robot, allowing one to easily study its behavior (mainly phototropism), locomotion, and mobility.

With the original solar-ball robot, one needs to use time-lapse photography to study these effects. The charging of the capacitor takes a few minutes, depending on the intensity of sunlight. When the electricity is discharged into the motor, the robot lurches forward a short distance. For example, 10 hours (h) of motion with the original solar ball can be compressed into a few minutes of study with this robot.

While this particular robot doesn't incorporate the electronics for an onboard power supply, it still uses a light trigger. The circuit shown in Fig 12.2 controls the power from the batteries to the gearbox motor. The circuit reads the level of illumination that the robot sees. If the light level is high enough, it turns on the motor to the gearbox. The trip level of the circuit is user adjustable using potentiometer V1.

## Gearbox

Before we get into the construction of the robot, let's first look at the gearbox (see Fig. 12.3). Physically, this gearbox is smaller than many gearboxes and is easier to fit inside the sphere. It has a 1000:1 gear ratio. The higher the gear ratio, the slower the robot will move.

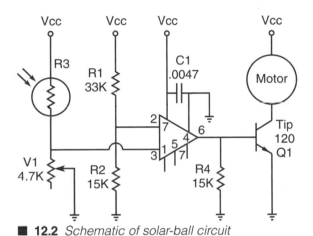

■ **12.2** *Schematic of solar-ball circuit*

■ **12.3** *Gearbox 1000:1 ratio*

In the prototype, the gearbox is set to the 1000:1 ratio. You can use any gearbox that can fit and rotate inside the 5$\frac{1}{2}$" transparent sphere. Choose one with a high gear ratio that delivers low revolutions per minute (7 rpm).

## Robot construction

The shell is the first component for consideration. It must be transparent and large enough to hold the gearbox and electronics. The shell used in my prototype has a diameter of 5$\frac{1}{2}$". Snap-together transparent spheres are available in many hobby and craft stores. Hobbyists use them to enclose holiday ornaments, If you cannot find a suitable shell locally, you can purchase one from Images SI (see the parts list at the end of this chapter). The plastic shell is fragile. Do not have your robot try to climb or fall down stairs; it is sure to crack and break.

Separate the two halves of the shell. The first job is to locate the center of each half sphere. This is where the shafts of the gearbox will be connected. Locating the center at first appears much easier than it actually is. To find the center, I was forced to trace the diameter of the shell on white paper, then draw a box around the drawn circle that touched the circle on four sides (see Fig. 12.4). Drawing

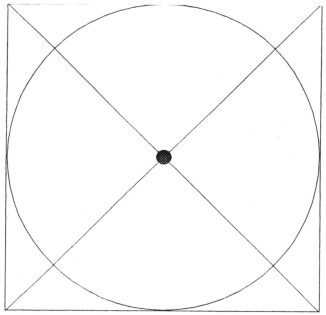

■ **12.4** *Layout for finding circle center*

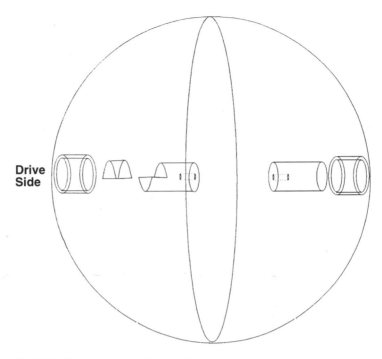

**Drive Side**

■ **12.5** *Transparent sphere drive components*

the diagonal lines from the corners of the box, I was able to locate the circle center. The half sphere is then positioned over the drawn circle. If you hold your head directly over the half sphere, you may be able to eyeball the center and mark it on the sphere with a magic marker. I tried this once or twice with less than ideal results. Finally, I taped the paper on a $\frac{1}{2}$" piece of wood and drilled a small hole at dead center. Then I placed a small dowel, about 2.5" long, in the hole, making sure it was perpendicular to the wood. Place the half sphere over the fixture, lining up its diameter with the drawn circle; the dowel locates the center of the sphere fairly accurately. Mark the center of one half sphere, and then the other.

The next step is to make a drive-locking fixture in the sphere that prevents the gearbox shaft from rotating freely inside. With the shaft locked, it forces the gearbox itself to rotate inside the sphere, changing the center of gravity and moving the robot along. The drive fixture must at the same time allow the sphere to be assembled or unassembled at will. The system I devised is illustrated in Figs. 12.5 and 12.6. Although I built all the drive components out of transparent plastic on the prototype, you can fabricate the parts out of other materials like brass and wood.

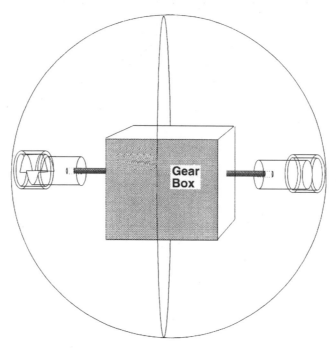

■ **12.6** *Gearbox placement in sphere*

The first component is a small length of tubing ⁵⁄₈" outside diameter (OD), ¹⁄₂" inside diameter (ID), and about ³⁄₈" long. This tubing is glued to the center of the half sphere, using the marks as a guide.

Inside the tubing, glue a ¹⁄₂"-diameter half round about ³⁄₈" long. This piece may be glued inside the tubing before the tubing itself is glued into the sphere.

Next, cut a small length of ¹⁄₂"-diameter solid rod. On one end of the rod, remove a ³⁄₈" half section. This is accomplished using a hacksaw or coping saw. First make a cut directly down the center of the rod about ³⁄₈" deep. Then make a horizontal cut to remove the half section. Check to make sure this shaft fits easily into the ³⁄₈" tube and half round assembly inside the half sphere. If not, file the cut end until it does. At the opposite end of this rod, drill a hole down the center that will fit the shaft from the gearbox.

*Note:* On the prototype robot, I made the second shaft a drive connection also. Only when the robot was finished did I realize that this was unnecessary. A single drive connection works just as well as a double.

The second half sphere is easier to make. Glue a small length of ⁵⁄₈"-OD, ¹⁄₂"-ID tubing to the center of the half sphere, using the mark as a guide. Cut a small length of $1/2$"-diameter solid rod. Check to make sure the shaft fits easily into the $5/8$" tubing. If not, obtain a small piece of 100-grit sandpaper. Wrap the sandpaper around a $1/2$" length on the end of the shaft. Twist the sandpaper around on the shaft to sand the end. Continue sanding until the end of the shaft fits easily into and out of the tubing. Next, drill a hole down the center that will fit the shaft from the gearbox.

We want the gearbox to be positioned in the center of the sphere. Place the shaft of the gearbox in the plastic rod. Place the rod in the tubing on half the sphere of the globe.

Position the gearbox so that it will lie in the center. Mark the depth the gearbox shaft must go into the plastic rod on the gearbox shaft. Remove the gearbox shaft. Mix a small amount of two-part epoxy glue. Coat the gearbox shaft with the epoxy glue and insert it into the plastic rod. Let the glue set before proceeding.

Once the glue has dried on the first shaft, we must glue the other plastic rod on the opposite side of the shaft. Position the glued rod into the half sphere. Place the other plastic rod on the opposite shaft. Place the other half sphere together with the first. Gauge the depth the gearbox shaft must be inserted in the plastic rod, and then add another $1/8$" of depth for error. Glue and let set. Check your work while the glue is setting on the second shaft to ensure that you can close the sphere properly.

# Electronics

The electronic circuit is a light-activated on/off switch. When the ambient light level is low (user adjustable), the circuit shuts off power to the gearbox. The user adjusts the sensitivity of the circuit using potentiometer V1.

There is nothing crucial about the circuit. If you do not wish to purchase or make the printed circuit board (PCB), the circuit may be wired and assembled on a standard breadboard.

## How it works

The circuit configures a complementary metal-oxide semiconductor (CMOS) operational amplifier (op-amp) as a voltage comparator. A comparator monitors two input voltages. One voltage is set up as a reference voltage called "Vref." The other voltage is the input voltage called "Vin," which is the voltage to be compared. When the Vin voltage falls above or below the Vref, the output of the comparator (pin 6) changes states.

The two input voltages are applied to pins 2 and 3. Pin 2 (inverting input) is connected to a reference voltage of approximately 1.5V, using a simple voltage divider made of resistors R1 and R2.

Photosensitive resistor R3 makes up another voltage divider in conjunction with potentiometer V1, which is connected to the noninverting input (pin 3) of the op-amp.

There is no feedback resistor between the output (pin 6) and either of the inputs (pins 2 and 3). This forces the op-amp to operate at its open loop gain.

A cadmium-sulfide (CdS) photoresistor is used as the light sensor. A photoresistor changes resistance in proportion to the intensity of the light that falls on its surface. The CdS cell produces its greatest resistance in total darkness. As the light intensity increases, its resistance decreases. In the circuit, the CdS cell is part of a voltage divider. The changing resistance of the CdS cell changes the voltage drop across the potentiometer V1, which is connected to pin 3. As the light intensity increases, the resistance of the CdS cell decreases, which increases the voltage drop across the potentiometer. This increased voltage drop is seen as a rising voltage. The trigger voltage can be set for different light levels using the potentiometer.

The electronic circuit is not crucial. You can construct the circuit using point-to-point soldering on a prototyping breadboard. A PCB is available from a kit, or you can make it yourself. The PCB artwork is illustrated in Fig. 12.7. Parts placement on the board is shown in Fig. 12.8.

Once the circuit is complete, you need to adjust the light level that will activate the circuit using potentiometer V1. Make temporary connections to the gearbox motor using alligator clip wires. Power to the circuit and gearbox is obtained from two AA cells, and the AA cell pack is glued to the back of the gearbox during final assembly. Make sure the battery pack has a battery clip for easily disconnecting and connecting power.

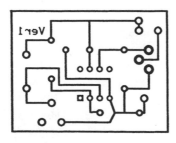

■ **12.7** *PCB layout*

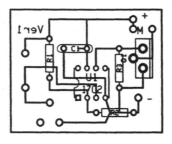

■ **12.8** *PCB parts placement*

When making the light-level adjustment, use a low level of light to activate the robot. When the robot is on the floor, if the light level is set too high it will stop every time it passes under a shadow.

## Putting it all together

Once the circuit is adjusted, you are ready for the final assembly. Glue the AA battery pack to the back of the gearbox, making sure that no glue comes into contact with any of the gears. Glue the electronic circuit board to the front of the gearbox, again making sure that none of the glue touches any of the gears. Connect the power supply. At this point the gearbox will probably start turning. To load the mechanism inside the robot, bring all the parts into a dark room to deactivate the circuit. Load the assembly inside the sphere.

Take the robot out into the light. The gearbox should become active. Place the robot on the floor. The robot should travel toward or in the direction of light. If the robot does the opposite, stop the robot, remove the gearbox and electronics, and reverse the wires leading to the motor.

## Locomotion

I was pleasantly surprised when I began observing this robot. I originally thought it would become trapped easily. Not so. When the robot enters a corner and stops, the gearbox inside begins swinging all the way up and over, radically shifting its weight over top dead center and moving the robot out of the corner.

## Advancing the design

When I originally designed this robot, I planned to use a steering mechanism to track a light source. However, the small steering mechanism didn't have enough weight to turn the robot in any direction quickly. In the long run, other factors (terrain, obstacles, etc.) affect its direction. Hence I removed the steering. But this is still a good research area for advancing the overall design.

### Adding higher behavior module

As the robot stands, when a certain level of light is reached it becomes active. We can add a higher behavior mode, feeding, by adding a few more components (two solar cells and steering diodes) and another comparator circuit. The second comparator circuit will deactivate the motor when the light illumination level becomes high enough, allowing the solar cells to charge the AA batteries, which will be changed to nickel-cadmium (NiCd) batteries.

Figure 12.9 illustrates the behavior. When the light level is low, the robot is off, or we can say it is in a resting mode. As illumination increases, it reaches a point where the motor turns on and the robot enters its searching mode. When the light level increases significantly beyond this point (searching mode), the second comparator turns off power to the gearbox motor, allowing the two solar cells to charge the AA NiCd batteries, which triggers the feeding mode.

If you plan to add this feeding behavior circuit, keep track of the current drain to the comparator circuits. It must not exceed the current supplied by the solar cells or obviously no charging to the NiCd batteries will occur.

## Parts list for the solar-ball robot

- ☐ (1) 5½" transparent plastic globe (see text earlier in this chapter)
- ☐ (1) Gearbox (see text earlier in this chapter)
- ☐ (1) 6" length of ½" solid plastic rod
- ☐ (1) 3" length of ⅝"-OD, ½"-ID plastic tubing
- ☐ (1) 1" length of ½" half-round plastic rod

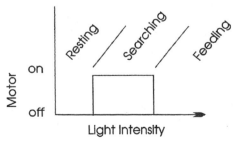

■ **12.9** *Layering higher behavior*

## Electronics

- ☐ (1) 5V CMOS op-amp ALD 1702 or equivalent
- ☐ (1) 33K-ohm, $\frac{1}{4}$-W resistor
- ☐ (1) CdS photoresistor
- ☐ (1) 4.7K-ohm potentiometer (PC mount)
- ☐ (2) 15K-ohm resistors
- ☐ (1) 0.0047-$\mu$F capacitor
- ☐ (1) TiP 120 NPN Darlington
- ☐ (1) PCB

A kit containing all the above components is $65.00. Shipping and handling via UPS Ground Service is $9.50. New York state residents add 8.25% sales tax.

Parts are available from:

Images SI, Inc.
39 Seneca Loop
Staten Island, NY 10314
(718) 698-8305

257

# Underwater bots

UNDERWATER ROBOTICS IS AN EXPANSIVE FIELD. MOST underwater robots are designed for salvage operations or exploration. In the future, underwater robots will help farm the ocean for fish, pharmaceuticals, minerals, and energy.

Underwater robots may also be used as mock-ups to test space-faring robots. A neutrally buoyant robot is essentially weightless. Propellers and motors replace rockets on these underwater robots. The lack of friction encountered in space can only be simulated in the underwater environment. If you want to design a robot that will function in space, a good place to start is with an underwater robot.

The National Aeronautics and Space Administration (NASA) has funded the development of telepresence remotely operated vehicles (TROVs) (see Fig. 13.1) and autonomous underwater vehicles (AUVs). The TROV tests virtual-reality (VR) based telerobotic techniques. Telepresence technologies are increasingly more important in exploration and hazardous duty. Telepresence technology will continue to grow in these fields and expand into others like entertainment.

## Dolphins and tunas

Interestingly, studies are being conducted that examine the swimming motion and propulsion of fish. It is common knowledge that underwater animals move and swim more efficiently than a ship's propeller can move a ship. Want to prove this to yourself easily? Have you ever tapped on the glass of an aquarium filled with fish? The sudden noise sometimes causes the fish to dart around so quickly your eyes can't follow their movement. Imagine if you

■ **13.1** *NASA TROV craft. Photo courtesy of NASA*

could design a ship that could move that fast, that suddenly. It's not surprising then that the U.S. government is funding some of these studies.

How efficient are fish at swimming compared to our current method of water propulsion? Let's glimpse at a partial analysis. In 1936 James Gray, a British zoologist, studied dolphins. His purpose was to calculate the power a dolphin needed to move itself at 20 knots, a speed at which dolphins are commonly reported to be able to swim. Gray's model of the dolphin was rigid, assuming that the water resistance for a moving dolphin is the same for a rigid model and flexible model. This is not true, but even accounting for this error, the conclusion Gray calculated is interesting. The dolphin is too weak, by a factor of 7, to attain the 20-knot speed. One may further deduce that the dolphin may be able to reduce its water resistance by a factor of 7 to compensate. But this probably isn't the entire answer either.

Well, for the last 60 years no one has been able to prove or disprove Gray's calculations conclusively. Any swimming mechanism that mimics fishlike swimming is grossly inefficient. Recently new studies are under way to again study fishlike swimming. With new computer technology behind this endeavor, scientists hope to answer these long-held questions.

Researchers at the Massachusetts Institute of Technology (MIT) in Cambridge have been studying the bluefin tuna for the last several years. They have created a 4-foot (ft) model "robot fish" that swims down the Ocean Engineering Test Tank Facility. The robot fish resembles a real fish. The skin is made of foam and Lycra. The robot uses six external motors that are connected to pulleys and tendons within the robot. The fish moves and swims like a real bluefin tuna.

## Swimming with foils

The tail of a fish is considered a hydrofoil. As the tail flaps side to side, it pushes water backwards and propels the fish forward. As the tail moves, vortexes are formed in the water behind it. It is believed that the vortex formation is key to understanding the greater efficiency of fish propulsion.

Dolphins are interesting; their hydrofoil tail lies horizontal. So instead of moving their tails side to side like fish, they move their tails up and down. This creates the same efficient thrust in water propelling the dolphin forward.

Penguins swim by using the thrust generated by their wings. Pictures of penguins swimming in water strongly resemble those of flying birds. There is a difference though. With birds in flight, the beating of their wings must supply lift as well as forward thrust. The lift is necessary to counteract the force of gravity. With penguins there is no necessity of lift. The density of water equals that of a penguin's body (neutral buoyancy), so the flapping of a penguin's wings simply needs to produce forward thrust.

## Paddles and rows

Since we're looking at methods of locomotion in water, we might as well include paddles and rows. Ducks use their webbed feet as paddles when swimming through water. Water beetles use their legs as oars and row themselves along like tiny boats.

## What have we learned so far?

Studies at MIT lead researchers to use a fluid dynamic parameter known as the Strouhal number. For fish, the number is calculated by multiplying the frequency of the tail flapping back and forth by the width of the vortex created in the water divided by the fish's speed. A number of species of fish were studied. The results were that maximum efficiency is found when the Strouhal number lies between 0.25 and 0.35.

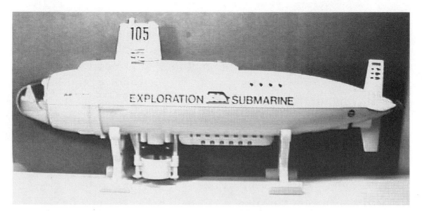

■ **13.2** *Toy submarine ready for conversion to TROV*

When the foils of the robot fish at MIT were adjusted and reconfigured to generate a Strouhal number in this range, its efficiency jumped higher than 86 percent. This is a major improvement compared to propellers that generate efficiencies around 40 percent.

## Jumping in

There are two basic underwater robot projects outlined in this chapter. One involves modifying a toy submarine, the other building a robotic fish from scratch.

### Submarine

There are a number of companies that make and sell hobby model submarines. Depending upon the degree of sophistication of the model, it usually is radio controlled (R/C) and is capable of submerging and surfacing (see Fig. 13.2).

In modifying a toy submarine, forget about R/C and jump to wire control and using an umbilical cord to the submarine. The umbilical cord can carry power as well as command and control signals.

These hobby submarines can be modified to create small telepresence systems. The primary modification is the addition of a color video camera. Most of these hobby submarines have open compartments where electronics gear can be stored (see Fig. 13.3).

Many of the systems used in the telerobotic car built in Chap. 9 can be implemented here. The one difference is the use of wire control instead of R/C.

Because these are "toy" subs, you will probably not let them loose in open water. The tiny propulsion motors in these submarines

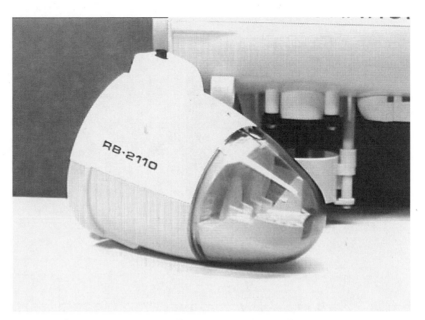

**■ 13.3** *Open compartment for housing electronics*

need calm water to function. Of course, these could be a starting point for more robust systems.

Are there any applications of these toy submarines beyond the experience gained from building and using basic underwater telepresence systems? I could imagine 10 or more toy telepresence submarines released in a swimming pool, each submarine being remotely controlled by one person. I'm sure a number of underwater or space scenarios could be created for game play.

## Swimming by use of a tail

As stated earlier, most mechanisms that mimic fishlike movement are grossly inefficient. This model is no exception. However, information gleaned from resources like MIT can be incorporated into the model (not done here) to improve overall efficiency. And if one plans on building like-animal androids, this is as good a place to start as any.

### Rotary solenoid

The robotic fish pivots on the use of a rotary solenoid (see Fig. 13.4). When activated, the solenoid rotates its top plate about 30 degrees. A spring returns the plate to its original position when deactivated.

The solenoid's top plate has at least two 3/48 threaded holes that may be used for mounting objects to it. The bottom of the solenoid

*Underwater bots*

■ **13.4** *Rotary solenoid*

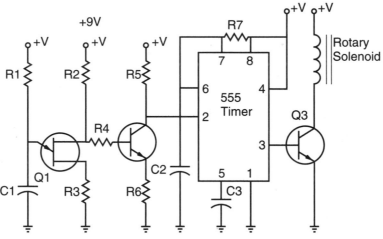

■ **13.5** *Schematic of electronics*

has two protruding 3/48 threaded rods that can be used for mounting the solenoid. The solenoid is not as powerful as I would like, but it is strong enough to provide underwater propulsion.

### Electronics

The electronic circuit uses a unijunction transistor Q1 (UJT 2646) to generate a slow stream of pulses (see Fig. 13.5). The timing of the pulses is determined by C1 and R1. The pulses pass through R4 to the base of Q2. Q2 is an NPN transistor 2N2222. The purpose of Q2 is to invert the pulse signal for input to IC1 pin 2. IC1 is a 555 timer configured in monostable mode. IC1 shapes the pulse width. The output of the 555 timers switches Q3 on and off. Q3 controls the current to the rotary solenoid that is used in the robotic fish.

The circuit is powered by a single 9-volt (9V) battery. The circuit is simple enough to hardwire on a prototyping style printed circuit board (PCB).

Test the circuit by connecting it to the rotary solenoid before continuing. The time period of the pulse should be approximately 1 second (s).

### Mechanics

To keep weight and mass down, most of the components are made out of aluminum. The first mechanism I used to convert the solenoid movement to a flapping fish tail is shown in Fig. 13.6. This turned out to be more complex than was necessary. Figure 13.7 shows the final tail assembly setup.

A $\frac{1}{8}$" thick × $\frac{1}{2}$" wide × $5\frac{1}{2}$" long piece of aluminum bar is secured to the top plate of the rotary solenoid using two 3/48 × $\frac{1}{4}$" screws. First drill two holes in the aluminum bar to match the holes in the top plate. Next, one hex nut is screwed flush to the underside of each screw head to prevent the screws being driven too far down. If the screws are driven too far down, they will prevent the top

■ **13.6** *Original tail assembly*

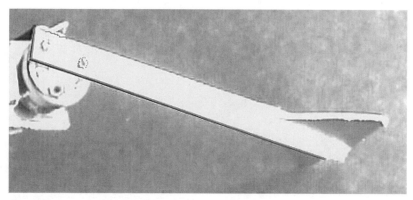

■ **13.7** *Final tail assembly*

plate from turning easily. Secure the aluminum bar to the top plate using the screws.

Fins are made by cutting a square of $1\frac{1}{4}$" aluminum diagonally. The fins to the tail are secured to the $\frac{1}{2}$" aluminum bar using a generous amount of hot glue. You may want to rough up the aluminum surfaces with sandpaper for a better bite before gluing.

The solenoid itself is secured to the end of a piece of aluminum $\frac{1}{8}$" thick $\times$ $1\frac{1}{4}$" wide $\times$ 2" long, using the two bottom 3/48 threaded studs and a few 3/48 hex nuts. The circuit and battery are secured on the front of the aluminum (See Fig. 13.8).

### Getting wet

Obviously we have an exposed circuit and solenoid. To prevent water from damaging any components, cover the components using a thin, transparent plastic sandwich bag. The bag is secured to the tail using wire. The bag should be such that the tail section can still move back and forth easily.

Before you dump the robot into the water, it should be made neutrally buoyant. If you dump it in as is, the front-heavy robot will nosedive to the bottom of the water tank with the tail swishing back and forth ineffectively. Secure strips of Styrofoam to the front of the model on the outside of the transparent bag using rubber bands. Place the model in water to test it. When the model submerges and floats underwater in a level or almost level position, you're ready to go. Turn on the power to the circuit and let the robot go.

### Efficiency

This particular robot doesn't move with the grace or efficiency of a real fish, but it does move. I think efficiency can be improved by cutting the $\frac{1}{2}$" $\times$ $5\frac{1}{2}$" aluminum tail bar in half and then securing the halves back together using 2" of spring. This spring will allow the tail section to bend and flex and should create better thrust.

■ **13.8** *Finished robot fish*

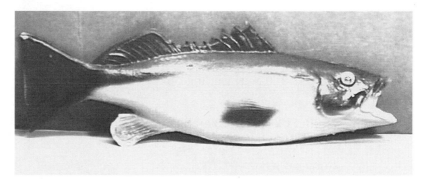

**13.9** *Rubber fish covering for robot fish*

### The robotic android fish

What separates a robot from an android is all in appearance. A robot looks like a robot and an android looks human or like the thing it was made to simulate. So an android fish should look like a fish.

Well, creating an android fish may not be as difficult as it may first appear. And the reason for this is that reasonable-quality fish coverings are available (see Fig. 13.9). These coverings are sold at magic and hobby shops. These rubber fish can be cut open and the robotic mechanism inserted inside.

Some models are more realistic than others. I found one model made out of thick, soft rubber. While the appearance and texture of this model is excellent, the internal robotic structure to move the tail must be more powerful. Another fish covering that is less realistic but much thinner, and therefore easier to move, is a better choice.

## Learn more about it

To learn more about fish-based propulsion systems, try reading the following sources: *Scientific American,* March 1995, "An Efficient Swimming Machine," by Micheal S. Triantafyllou and George S. Triantafyllou, and *Exploring Biomechanics,* by R. McNeill Alexander, published by Scientific American Library, 1992, ISBN 0-7167-5035-X.

## Parts list for robotic fish

☐ R1, 33K ohms
☐ R2 and R6, 100 ohms
☐ R3, 470 ohms

- [ ] R5, 10K ohms
- [ ] R7, 15K ohms
- [ ] Q2, NPN 2N2222 transistor
- [ ] Q3, TIP 120 NPN Darlington
- [ ] IC1, 555 timer
- [ ] C1 and C2, 22-microfarad ($\mu$F) capacitors
- [ ] C3, 0.01-$\mu$F capacitor
- [ ] Rotary solenoid—$5.95
- [ ] Q1 2N2646 UJT—$5.95
- [ ] $1/_8$" thick $\times$ $1/_2$" wide $\times$ 6" long aluminum bar—$1.50
- [ ] $1/_8$" thick $\times$ $1^1/_4$" wide $\times$ 2" long aluminum bar—$1.00

Parts are available from:

Images SI, Inc.
39 Seneca Loop
Staten Island, NY 10314
(718) 698-8305

http://www.imagesco.com

Shipping and handling via UPS Ground—$9.50

# 14

# Aerobots

AEROBOTS (AERIAL ROBOTS) ARE A CLASS OF ROBOTS THAT can fly. They include lighter-than-air aircraft (blimps), helicopters, and airplanes. Some applications for aerobotics are autonomous flight, drones, warfare, surveillance, advertising, and telepresence.

Autonomous aircraft have a long history, the first ones being built in the early 1920s. One unmanned aerial vehicle (UAV) code-named the "bug" was designed for warfare. The bug was about 12 feet (ft) long with a wingspan of 15 ft. Its sophisticated flight control system (for its time) included a gyroscope, an altimeter, and electric and pneumatic controls. The flight control system flew the craft 30 to 40 miles into enemy territory. When the desired distance was reached, the craft would jettison its wings, forcing the nose-heavy fuselage to fall to Earth carrying a payload of 200 pounds (lb) of explosives. But World War I ended before the bug could see any action.

From this beginning UAVs have been under continual development and refinement. The latest UAVs saw action in the Persian Gulf war. Although the UAVs received little to no press, they flew over 300 sorties. They performed reconnaissance and damage assessment and followed enemy weapons deployment. The most recognized and most sophisticated autonomous aircraft is epitomized by the self-guided cruise missile carrying nuclear warheads.

Telepresence flight control systems also have a long history, but not as long as that of UAVs. In World War II, the United States used remotely piloted aircraft to fly kamikaze missions. The old-style remote control systems have nowhere near the technical sophistication of today's systems. The old remote control systems were unreliable, and the pilot needed to keep a visual eye on the remote aircraft to fly it accurately.

Today remotely piloted aircraft have video cameras transmitting pictures back to the pilot. The pilot may be situated anywhere in the world. The systems have developed into telepresence virtual-reality (VR) systems.

The aerobot we will build is a flying telepresence blimp. The reason I chose this mode of flight (a lighter-than-air framework) over a model helicopter or model airplane is safety, silence, low cost, and ease of use.

Blimps are quiet, slow, graceful, and forgiving in flying errors. Safety was the major factor in my decision. If a blimp bumps into a person or object, there will be little or no harm. Airplanes and helicopters on the other hand are potentially lethal weapons (the propellers on airplanes and helicopters) when in close proximity to human life.

The blimp we will build is limited to indoor use. Care must be taken in choosing components that are extremely lightweight. The payload capacity (lift) of the blimp is approximately 6 ounces (oz). This means our radio-controlled (R/C) receiver, propulsion, power supply, charged coupled device (CCD) camera, and video transmitter together must weigh in at or under this 6-oz weight restriction. Tough, but not impossible.

## Lighter-than-air aircraft background

Lighter-than-air aircraft fall into three categories: rigid, semirigid, and nonrigid. Rigid aircraft have internal frames usually made from lightweight aluminum. The most famous of these are the Zeppelins.

Semirigid aircraft have a rigid lower keel section. A nonrigid envelope that is filled with helium is secured above it.

Nonrigid aircraft are the ones we are most familiar with today. These are blimps. One of the more famous blimps is the Goodyear blimp used for advertising. Nonrigid aircraft are made of huge gas envelopes. The envelope shape develops when it is filled with helium gas.

## Blimp systems

The most widely known use for blimps today is for a bird's eye view of major football games. Another popular use people are familiar with is for advertisements.

While blimps may seem like old technology, scientists and engineers are still developing uses for them. For instance, the U.S. Army has used an unmanned airship called SASS LITE (Small Airship Surveillance System, Low Intensity Target Exploitation). The SASS LITE is used for border patrols. Recently the manufacturer stated that this 90-ft airship is available for commercial ventures.

Helium balloons are capable of reaching the upper stratosphere. One company has proposed building an air station 100,000 ft above the Earth. The station would provide a telecommunications link just like a satellite. However, the air station would cost 50 percent less than a similarly equipped satellite.

Robotic systems and telepresence systems have been put on model blimps for a number of years. We will review two ventures shortly, one from the Robot Group and the other from Berkeley's WEB Blimp. What we will focus upon accomplishing is placing a simple telepresence system on a model blimp. In reality the telepresence system is a wireless, flyweight, portable surveillance system. Sensor feedback systems that could relay a sense of touch, for a "real" telepresence, are not developed. Our simple system transmits video and sound. The user or operator can move (fly) the blimp via radio controls.

## The Robot Group—Austin, Texas

Robotic systems have been placed on model blimps. The Robot Group, based in Austin, Texas, exhibited a robotic blimp at Robofest 1 in the fall of 1989. I'm sure robotic systems have been in place on blimps before this for military and scientific purposes; however, the Robot Group represents private (nongovernment funded) exploration in this area. The Robot Group continues to develop and improve upon the robotic blimp. In 1991 the computer blimp project called the Mark III used ultrasonic sensors and a neural network navigation system. Although the system fell short of design expectations, it did function properly.

The Robot Group has a website on the Internet which you can visit to get the latest information (see Internet Access at the end of this chapter).

## WEB Blimp—University of California, Berkeley

Space browser is the name given to telepresence blimp systems being designed and built at the University of California, Berkeley, Department of Electrical Engineering and Computer Science. The blimps are used as avatars, or as I prefer to call them, golems.

The Berkeley group is striving for tele-embodiment systems. A true tele-embodiment system would require a complex sensor feedback system from the blimp avatar to the user. Currently the feedback system provides video and sound. The user can maneuver the blimp via radio control.

The most interesting aspect of this blimp is that it may be controlled over the Internet, hence the name WEB Blimp. The video is fed to the Internet via a video frame grabber with a CU-SeeMe format output. The WEB Blimp is made available through the Berkeley website (see Internet Access at the end of this chapter).

## Designing telepresence blimps as avatars and golems

Almost as good as being there! Robotic blimps or a reasonable facsimile have a good future in the telepresence industry. Suppose you wanted to look at some paintings in the Louvre in Paris, visit the American Museum of Natural History in New York City, then jump over to the Smithsonian in Washington, D.C., and finally check out the penguins on the Galapagos Islands. And let's say you wanted to do all this in a couple of hours.

One way this may be accomplished in real time is through the use of telepresence systems. One day in the future there will be sightseeing telerobots you may jump into through a phone (or satellite) link and your home computer VR system. These robots will be located at many points of interest throughout the world.

The telerobots are not restricted to Earth. There will be telerobots in space, underwater, and flying through the air. The Jason project is one underwater science adventure for schools. Through a satellite link, schools set up a communication link to scientists on a remote vessel. Students are able to learn what the scientists are doing, ask questions, and sometimes operate a TROV (telepresence remotely operated vehicle) via the satellite link.

### To the moon

Lunacorp in Fairfax, Virginia, has plans to place a civilian rover on the moon (see Fig. 14.1). For part of the time the rover will be used as a telepresence system for earth-bound drivers (see Fig. 14.2). Unfortunately, the operation cost is expensive, approximately $7000 per hour. I don't know about you, but I'll plunk down $120 to drive a telepresence rover across the lunar surface for a minute.

Lunacorp plans to get the rover to the moon in 2003: the site, Tranquility Base. But we are digressing from our main topic of blimps.

■ **14.1** *Lunacorp rover*

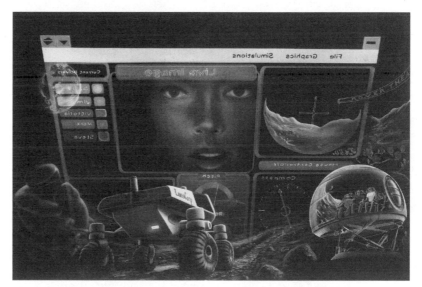

■ **14.2** *Artist's conception of Lunacorp telepresence system*

## Blimp parameters

Blimps need to meet certain design criteria to be used effectively for Earth-bound terrestrial telepresence systems. The blimps must be completely safe around human beings. The blimps should be able to move through the same passageways used by people. The CCD camera transmitting video should be positioned at approximately eye level. It should be able to navigate through gentle crosswinds without difficulty.

A ballast system needs to be created that would allow the blimp to achieve neutral buoyancy on or through a number of floors in a building. If a ballast system is too difficult to implement, neutrally buoyant robot blimps can be positioned on each floor. Operators would simply switch to an unused telepresence robot held on a requested floor when wanting to change floors.

Because of the low weight of blimps, they have physical restrictions. For instance, a blimp would not be able to push open a door. Buildings would need to be modified so that doors and elevators could be operated electronically using remote control signals emitted from the blimp.

## The blimp kit

The blimp we will construct is made of a tough Mylar material. The material can be heat sealed using a household iron. There are many different styles of blimps one may build: a flying-saucer shaped, delta-wing-glider shaped, or a typical "Goodyear" blimp. I recommend the simplest one of all, a pillow-shaped blimp.

Making a pillow-shaped blimp is easy. Fold the sheet of Mylar material in half (shiny side out). Heat seal the three open sides closed, leaving a little space that isn't heat sealed for a fill tube at the bottom, and you're finished.

### Helium

Helium is sold in canisters from many party stores to fill balloons. The canisters resemble those used to hold propane gas. If you don't have a local party goods store, look under helium or gas in the yellow pages to find a supplier.

### Helium versus hydrogen

When I first began this project, I thought about using hydrogen instead of helium, reasoning that because hydrogen weighs about half of what helium weighs, I could increase my lift by a factor of 2. Right? Wrong!

While I was correct in my assumption that hydrogen weighs about half of what helium weighs (see Table 14.1), I was incorrect in calculating the lift. Here's why. Lift is generated by the amount of air displaced by the helium (or hydrogen), just like an air bubble in water. Let's use this analogy. The air is less dense than the surrounding water, so the air bubble rises to the surface. Likewise, helium is less dense than the surrounding air; therefore, it rises also. Think of the rising helium or hydrogen as floating on top of a much denser gas we call air (see Table 14.1).

|                     | English, lb/ft$^3$ | Metric, kilograms per cubic meter (kg/m$^3$) |
|---------------------|-----------|-----------|
| Weight of hydrogen  | 0.0058    | 0.09      |
| Weight of helium    | 0.0110    | 0.178     |
| Weight of air       | 0.0807    | 1.29      |

So what's the lift of a helium balloon with 5 ft$^3$ of displacement?

Weight of displaced air = 5 (0.0807) = 0.4035 lb

Weight of 5 ft$^3$ of helium = 5 (0.0110) = 0.0550 lb

Lift = 0.4035 lb − 0.0550 lb = 0.3485 lb

That's quite a bit of lift! The reason is that we didn't subtract any weight for the balloon. If the balloon weighs 0.25 lb, the usable lift (0.3485 lb − 0.25 lb) is reduced to 0.0985 lb, or 1.57 oz.

How does this compare to the lift using hydrogen? Well, the weight of the air displaced is the same.

Weight of 5 ft$^3$ of hydrogen = 5 (0.0058) = 0.029 lb

Lift = 0.4035 lb − 0.029 lb = 0.3745 lb

The difference in lift for a 5-ft$^3$ balloon is

0.3745 lb − 0.3485 lb = 0.026 lb, about $^1\!/_2$ oz

Because the difference in lift is small, it is not worth the added risk of using hydrogen gas! I recommend using helium gas only.

## Size

The piece of Mylar used to make a balloon after it's folded in half, lying down flat, measures 34" × 56". The weight of the material is 3 oz (0.1875 lb). It's difficult to estimate how much helium the balloon will hold. To make a rough estimate, I use the volume of a cylinder. I know a pillow shape is not a cylinder, but, like I said, it's a rough estimate. First find the diameter. The material is 34" × 2 equaling 68" for the circumference. The circumference of a circle is 2 times pi (3.14) times the radius. If you do the math, the radius works out to 11". The volume of a cylinder equals pi times the radius squared times the height. The height in this case is 56". If you do the math, the volume equals about 12 ft$^3$.

The balloon will not be filled to its maximum capacity. In this case I'd estimate the balloon will hold about 70 percent of the calculated volume or about 8.4 ft$^3$ of helium gas.

275

### Calculated lift

$$\text{Weight of air} = 8.4 \ (0.0807 \ \text{lb/ft}^3) = 0.678 \ \text{lb}$$
$$\text{Weight of helium} = 8.4 \ (0.0110 \ \text{lb/ft}^{3)}) = 0.0924 \ \text{lb}$$
$$\text{Weight of Mylar material} = 3 \ \text{oz, or } 0.1875 \ \text{lb}$$
$$\text{Lift} = + \ 0.678 - 0.0924 - 0.1875 = 0.398 \ \text{lb,}$$
$$\text{or } 6.37 \ \text{oz}$$

## Construction

Construction of the blimp is simple and straightforward. Essential to the construction is being able to make a good heat seal. Cut a small section of Mylar material from the large sheet to practice on. Fold the small section of Mylar, shiny sides to the outside, dull sides together. Set the iron to a medium heat setting. Keep adjusting and testing the heat setting of the iron on small Mylar practice scraps until you find the right temperature. Every time you adjust the temperature of the iron, allow at least 5 minutes for the iron to stabilize to the new temperature. If the temperature setting is too hot, the Mylar material will melt and create holes. If the temperature setting is too cold, the heat seal will pull apart too easily. A good heat seal will not pull apart easily. Allow the Mylar material to cool for a minute before you test the heat-sealed seam. Once you have found the right temperature setting, write it down for future reference.

We will make a pillow-shaped blimp. Fold the sheet of Mylar material in half (shiny side out). Heat seal the three open sides closed, leaving a little space that isn't heat sealed for a fill tube at the bottom, and you're finished. The heat-sealed seam should be $\frac{1}{2}$" to 1" wide.

## CCD camera

The CCD camera provides the video from the blimp (see Fig. 14.3). Naturally, weight is a consideration. This camera weighs a little over half an ounce. The overall size is $1\frac{1}{4}$" $\times$ $1\frac{1}{4}$" $\times$ $1\frac{1}{8}$". Light sensitivity is 0.03 lux. Resolution is 430 TV lines. Output video is a standard NTSC (1V pp) signal. The camera can be powered from 9 to 12 volts direct current (VDC) maximum. The current draw from the camera is approximately 100 milliamps (mA).

A 9V transistor battery can power the camera. The battery weight (1.5 oz) is three times greater than the weight of the camera itself.

## TV transmitter

There are a number of TV transmitter kits available. There are two basic classes of transmitters. One type transmits the video and audio

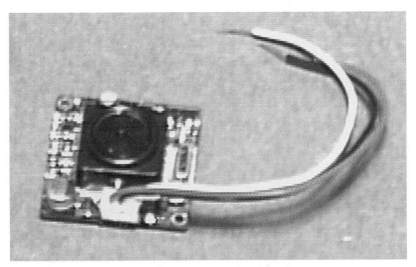

■ **14.3** *Lightweight CCD camera for telepresence system*

on one of the standard TV channels. The TV's tuner picks up the signal and displays it. These transmitters have a limited range of a few hundred feet.

The second type is more expensive. This type transmits well above TV frequencies into the 900-megahertz (MHz) range, and the TV requires a down converter to display the video. The down converter receives the 900-MHz signal and down converts it to a standard TV frequency. These units have a much greater range and better fidelity. The unit used in this prototype transmits directly to a TV set on its VHF (Channel 14) channels (see Fig. 14.4).

## Radio-control system

The radio-control (R/C) system is specially designed for blimps (see Fig. 14.5). It is extremely lightweight. The propulsion unit is a twin turbo fan that attaches to the underside of the blimp. Each turbo fan is bidirectional, and each is controlled by its own channel on the two-channel transmitter.

This helps the maneuverability of the blimp. While one turbo fan pushes forward, the other pushes backward, helping turn the blimp quickly. The pillow blimp ready for the telepresence system is shown in Fig. 14.6. A close-up of the turbo fan, miniature CCD camera, and TV transmitter is shown in Fig. 14.7.

### Going further

The blimp as it stands is a telepresence system. By placing autonomous navigation in it, we can convert the blimp into a flying robot.

■ **14.4** *TV transmitter circuit*

■ **14.5** *Lightweight R/C control system for blimp*

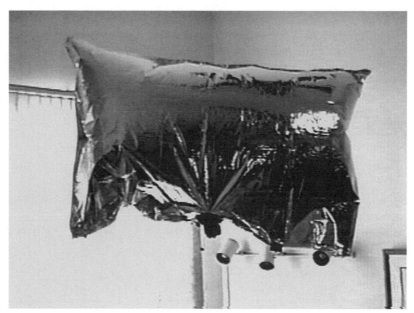

■ **14.6** *Pillow blimp*

■ **14.7** *A close-up of the turbo fan, miniature CCD camera, and TV transmitter*

# Parts list for the blimp

☐ (1) Blimp with radio controls, #T30824-77—$79.95

Part is available from:

Edmund Scientific
60 Pearce Ave
Tonawanda, NY 14150-6711
1-800-728-6999

☐ (1) Mini B/W CCD camera—$64.95

☐ (1) Mini TV transmitter—$90.00

Parts are available from:

Images SI, Inc.
39 Seneca Loop
Staten Island NY 10314
(718) 698-8305
http://www.imagesco.com

# Internet access

☐ Robot Group, Austin, Texas—Neural Net Blimp
http://www.robotgroup.org/projects/mark4.html

☐ WEB-controlled blimp at Berkeley
http://vive.cs.berkeley.edu/blimp/

☐ WEB Blimp
http://register.cnet.com/content/features/quick/weblimp
http://utopia.minitel.fr/~mpj/airships/">Marv's Airship Server

☐ University of Virginia—Solar-powered airship
http://minerva.acc.Virginia.edu:80/~secap/

☐ U.S. competitor in Australia Solar Challenge
http://www.mane.virginia.edu/airship.htm

☐ Intelligent surveillance blimp at the University of Virginia
http://watt.seas.virginia.edu/~jap6y/isb/

☐ Japanese Project—Solar-powered airship
http://www.aist.go.jp/mel/mainlab/joho/joh04e.html

# 15

# Robotic arm and IBM PC interface and speech control

THIS IS A MULTILEVEL MODULE PROJECT. THE FIRST MODULE is a robotic arm that is purchased in kit form. The second module is the IBM personal computer (PC) interface kit. The third module is a speech-control module.

The robotic arm may be operated manually with a manual control box that comes with the robotic arm kit. The robotic arm will work with either the IBM PC interface kit or the speech-control module. The IBM PC interface kit allows the robot to be controlled and programmed via a host IBM PC computer. The speech control allows you to operate the robotic arm via voice.

Together these modules form a functional unit that permits you to experiment and program automation and animatronics into a fully "wired" robotic arm.

The PC interface allows you to use your personal computer to program automation and animatronics into a robotic arm. You also have the option to control the arm interactively using either a manual controller or the Windows 95/98 program. Animatronics is the entertainment side of automation. For instance, if you covered the robotic arm with a child's sock puppet and programmed a small show, you would be programming an animatronic or electronic puppet. Programming automation has widespread applications in industry and entertainment.

The most widely used industrial robot is the robotic arm. Robotic arms are extremely versatile, due to the fact that the end manipulator of a robotic arm can be changed to fit particular tasks or industries. For instance, welding gear manipulators are used in spot-welding robots, spray nozzles for spray painting parts and assemblies, and grippers for pick and place, to name a few.

So you see, robotic arms are useful and make for an ideal learning tool. However, building a robotic arm from scratch is a difficult task. It is far easier to assemble a robotic arm from a kit. OWI sells a suitable robotic arm kit available from a number of electronic distributors (see the parts listing at the end of the chapter). The interface connects the finished robotic arm kit to a host computer printer port. The host computer is any IBM PC or compatible computer capable of running DOS or Windows 95/98.

Once connected to the computer's printer port, the robotic arm may be operated interactively and programmed from the computer. Operating the robotic arm interactively is easy. Simply click on any function button to command the robotic arm to perform that function. Click on the button a second time to end the function.

Programming automation is just as easy. First click on the Program button to enter the program mode. In this mode, the interface and

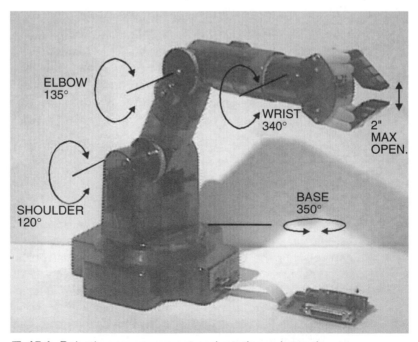

■ **15.1** *Robotic arm movement and rotation schematic*

robotic arm operate as described before, but in addition, each function and the time elapsed are recorded into a script file. The script file can contain up to 99 separate functions, including pauses. The script file itself can be replayed 99 times. Writing different script files allows one to experiment with computer-controlled automation and animatronics. The Windows 95/98 program operation is described in more detail later on. The Windows program is included in the robotic arm interface kit or may be downloaded at no cost from the Internet at http://www.imagesco.com.

In addition to the Windows program, the arm can be operated using BASIC or QBASIC. A DOS-level program is included on the disks that come with the interface kit. However, the DOS program only allows interactive function using the computer keyboard (see BASIC listing on one of the disks). Script file creation is not available in the DOS-level program. However, if one knows how to program in BASIC, the arm may be programmed to perform a sequence of motions similar to the script files created in the Windows program. The motion sequence may be repeated, as is done in many animatronics.

## Robotic arm

The robotic arm (see Fig. 15.1) can move freely in three axes of motion. The elbow joint can move vertically (up or down) through an arc of approximately 135 degrees. The shoulder joint moves the gripper forward and back through a 120 degree arc. The arm can turn clockwise (CW) or counterclockwise (CCW) from the base approximately 350 degrees. The gripper portion of the robotic arm can grasp and release small objects up to 2" in diameter and finally can rotate the gripper section approximately 340 degrees at the wrist.

The OWI Robotic Arm Trainer uses five small direct current (DC) motors to produce motion. The motors provide a "wire" control to the robotic arm. "Wire control" means that each robotic function (and hence DC motor) is controlled by a wire (electric power). Each of the five DC motors controls a robotic arm function. The wire control makes it possible to build a controller unit for the arm that will respond to electrical signals. This simplifies the task of interfacing the robotic arm to a PC printer port.

The arm is made from lightweight plastic. Most of the stress-bearing parts are also made of plastic. The DC motors used in the robotic arm are small, high revolutions per minute (rpm), low-torque motors. To increase the motor's torque, each motor is connected to a gearbox. The motor-gearbox assemblies are used inside the construction of the robotic arm. While the gearbox increases the motor's torque, the robotic arm is not capable of lifting or moving a great amount of weight. The maximum recommended lifting capacity is 4.6 ounces (oz) [130 grams (g)].

The robotic arm kit components have been thoughtfully laid out for kit builders (see Fig. 15.2). If you carefully follow the directions in the robotic arm construction booklet, construction will proceed smoothly. To help you, some of the assembly work is already completed. For instance, the five DC motors come with the gearboxes already assembled and connected to the DC motors (see Fig. 15.3). This helps move the construction along. In a few hours, you have an operational robotic arm.

## Basic motor control

To understand the basic function of wire control, let's see how digital signals can control a single DC motor. Controlling a DC motor requires two complementary transistors. One transistor is a PNP type and the other is an NPN type. Each transistor functions like a switch, controlling the current to the DC motor. The current direction controlled by each transistor is opposite to that of the other transistor. The direction of the current controls the direction the motor spins, CW or CCW. Figure 15.4 is a test circuit that you may build before building the robotic arm interface. Notice that if both transistors are turned off, the motor is off. Only

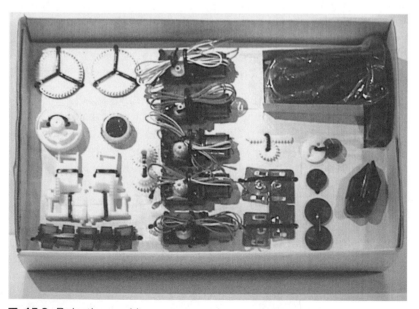

■ **15.2** *Robotic arm kit*

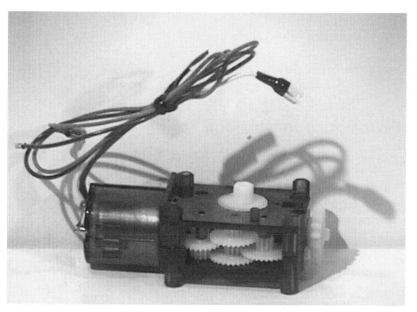

■ **15.3** *Preassembled gearbox*

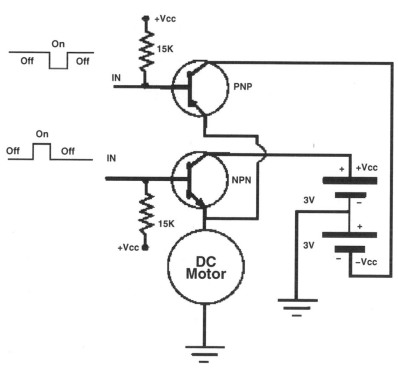

■ **15.4** *Test circuit schematic*

---

one transistor (per motor) should be turned on at a time. If two transistors to the same motor are accidentally turned on at the same time, it will be the equivalent of creating a short circuit. Each DC motor in the robotic arm is controlled by two transistors on the interface in a similar manner.

## PC interface construction

The PC interface schematic is shown in Fig. 15.5. If you purchase the PC interface kit, the printed circuit board (PCB) parts placement is shown in Fig. 15.6.

Begin construction by first identifying the component mounting side of the PCB. The component side has the white line drawings of the resistors, transistors, diodes, integrated circuits (IC), and DB25 connector. All components are mounted on the component side.

In general, after soldering a component to the board, clip away any excess wire from the underside of the PCB. It's a good idea to follow the sequence for mounting the components. Begin by mounting the

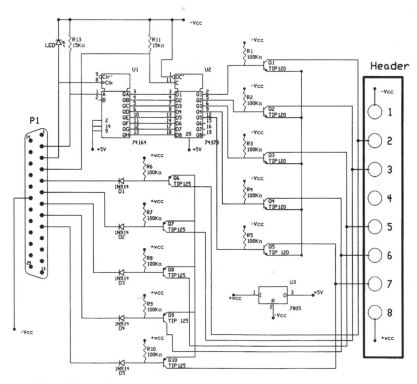

■ **15.5** *PC interface schematic*

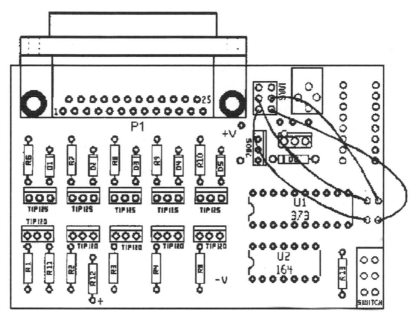

■ **15.6** *Parts placement diagram for PC interface*

100K-ohm resistors (color bands brown, black, yellow, gold, or silver) labeled R1 through R10. Next mount the fives diodes D1 to D5, making sure the black band on the diode faces toward the DB25 connector, as shown on the white line component drawing. Next mount 15K-ohm resistors (color bands brown, green, orange, gold, or silver) R11 and R13. Mount the red light-emitting diode (LED) in the R12 position on the board. The positive lead of the LED is mounted in the + label R12 hole. Next mount the 14- and 20-pin sockets in the U1 and U2 positions. Mount and solder the DB25 right angle connector. Do not force the DB25 pins through the board; it is a precision fit. If necessary, gently rock the connector in, making sure not to bend any pins. Mount the slide switch and the 7805 voltage regulator. Cut and solder four wires above the switch. Take care to keep the wire orientation as shown. Mount and solder the TIP 120 and TIP 125 transistors. Finish the project by mounting the eight-position header and 3" connection cable. The header is mounted so the longer leads face upward. Insert the two ICs, the 74LS373 and 74LS164, into their respective IC sockets. Be sure to orient the chip indentation on the top side of the chip with the indentation on the white line drawing. You may notice that there are places for additional components. This is for an additional AC adapter. Figure 15.7 shows the top side of the finished interface.

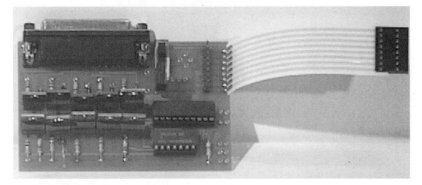

■ **15.7** *Assembled PC interface (top)*

## How the interface works

The robotic arm contains five DC motors. We need 10 input/output (I/O) lines to control each motor and direction. The parallel (printer) port on the IBM PC and compatibles has only eight I/O lines. To increase the number of I/O lines, the robotic arm interface incorporates a serial in, parallel out (SIPO) integrated circuit, the 74LS164. By using just two lines off the parallel port, lines D0 and D1, to send serial information to the chip, we can add eight I/O lines. Although eight I/O lines are available off the 74LS164, the interface requires just five of those I/O lines.

When serial information is transmitted into the 74LS164 chip, the parallel output of the chip shifts in response. If the outputs of the 74LS164 were directly connected to the transistors, the arm functions would switch on and off as the serial information clocked in. Obviously that would not be a suitable situation. To prevent this from happening, a second chip is added to the interface, the 74LS373 octal latch.

The 74LS373 octal latch has eight input lines and eight output lines. Binary information placed on the input lines is transmitted (made transparent) to the output lines when the chip is enabled. When the chip is not enabled, the information on the output lines is latched. When latched, the binary information on the input lines has no effect on the status of the output lines.

When all serial information has been transmitted into the chip, the 74LS373 octal latch is enabled, via parallel port pin D2. This allows the parallel information from the 74LS164 to be transmitted to the output lines of the 74LS373. The output lines from the 74LS373 switch the TIP 120 transistors on and off, thereby controlling the robotic arm functions. The process is repeated for each new command to the robotic arm. Parallel port lines D3 through D7 control the TIP 125 transistors directly.

## Connecting the interface to the robotic arm

The robotic arm uses a single 6V power supply consisting of four D cell batteries in the base. The PC interface takes power from the arm's 6V power supply. The power supply is used like a bipolar ±3V power supply. Power is tapped from the eight-conductor Molex connector to the arm base.

Connect the interface to the robotic arm using the 3"-long eight-conductor Molex cable. The Molex cable connects to the connector on the base of the robotic arm (see Fig. 15.8). Make sure the Molex connector is firmly and properly seated. To connect the interface to the computer's printer port, use the 6-ft DB25 cable supplied with the kit. One end of the cable connects to the computer's printer port. The other end connects to the DB25 connector on the interface board.

In most cases the printer port is also used for the printer. To alleviate switching cables back and forth whenever you want to use the robotic arm, purchase an A/B data switch (DB25) box. Connect the robotic arm interface printer to the A side and the printer to the B side. Now you can use the switch to connect the computer to either the interface or printer.

## Installing the Windows 95 program

Insert the 3.5" diskette labeled "Disk 1" into the computer's floppy drive and run the setup program (setup.exe). The setup program

■ **15.8** *Connecting the PC interface to the robotic arm*

creates a directory named "Images" on the computer's hard drive and the needed files are copied into this directory. An Images icon is created on the Start menu. To run the program, click on this icon in the Start menu.

## Using the Windows 95 program

Connect the computer's printer port to the interface using the 6-ft DB25 cable. Connect the interface to the base of the robotic arm. Keep the interface off for the time being. If you turned the interface on at this point, the existing information (status) left on the printer port may cause the robotic arm to begin performing a function.

Start the program by double clicking the Images icon in the Start menu. The program's opening screen is shown in Fig. 15.9. With the program running, the red LED on the interface should be blinking. *Note:* The interface does not have to be turned on for the LED to blink. How fast the LED will blink depends on the processor speed in your computer. The blinking light from the LED may be very dim; you may need to block some room light to see it by cupping your hands around the LED. If the LED is not blinking, the program is probably set to the wrong printer port address (LPT port). To set the interface to a different printer port address (LPT port), go to the Printer Port Options box on the upper right-hand corner of the

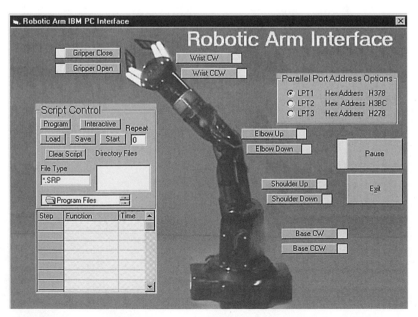

■ **15.9** *Screen shot of Windows PC interface program*

screen. Click a different option. Whichever port setting option causes the LED to start blinking is the correct printer port address.

With the LED blinking, click on the Pause button and then turn the interface on. Clicking on a function button causes the robotic arm to perform the corresponding function. Clicking on the button a second time stops that function. Using the function buttons to control the robotic arm in this manner is called the *interactive mode*.

### Creating script files

To program motion and automation into the robotic arm, we use script files. The script file contains a list of timed instructions that control the robotic arm function. Creating script files is simple. To create a script file, click on the program button. This puts you in the "programming" script writing mode. Clicking on a function button will start the robotic arm function as before, but in addition, the function information is entered into the yellow script table on the lower left of the screen. The step number is placed in the left column, beginning with step 1, and increments with each new function. The function name is entered in the middle column. When the function button is clicked the second time, the function stops as before, but the elapsed time from starting to stopping the function is entered into the third column. The time elapsed is given in increments of a quarter second. Continuing in this manner a user may program up to 99 functions, including timed pauses, into a script file. Script files may be saved to and loaded from the local directory. Script files may be set to replay up to 99 times by typing a number in the repeat box and hitting Start. To stop writing into a script file, click on the interactive button. This puts the computer back into the interactive mode.

# Animatronics

Script files may be used for computer automation or animatronics. With animatronics the underlying mechanical robotic system is usually covered and hidden from sight. Remember the sock puppet at the beginning of this chapter? The coverings vary from humans (whole or partial), aliens, animals, plants, or minerals to anything in between.

# Limitations

If you were performing automation or animatronics on a professional level, your robot would be required to be in the exact position needed, to hit its mark, so to speak, 100 percent of the time.

You will notice that as a sequence (script file) is continually repeated, the position of the robotic arm will drift from its original position. There are a number of reasons for this. As the battery power supply to the robotic arm becomes depleted, the reduction in electric power delivered to the DC motor reduces the torque and speed of the motor. So during a timed function, the motors will not move as far or lift as much with old batteries as they would with fresh batteries. But this isn't the entire case. Even with a regulated power supply, how many times the DC motor shaft spins in a given length of time is neither counted nor controlled. So the number of turns the motor spins in each timed sequence will vary by a small percentage. This causes the position of the robotic arm to drift. If that wasn't enough, the gears used in all the motor gearboxes have a certain amount of slop (or play) that isn't taken into account. All these factors taken together go a long way in explaining why the position of the robotic arm repeatedly performing a script file will drift over time.

## Finding home

To enhance this project, positional feedback from the robotic arm could be implemented so the computer could determine absolute position of the arm. With basic positional feedback, the robotic arm can be located in precisely the same position every time at the beginning of a script file (sequence) run.

There are many approaches one can take. One basic method doesn't employ positional control but instead uses limit switches to find a "home" or starting position. The limit switches determine when the arm reaches one absolute, or home, position. To accomplish this, a series of limit switches (momentary contact lever switches) will close when the arm reaches its limit of travel in that particular direction. For instance, one limit switch would be mounted to the base. This switch would close only when the robotic arm was turned completely in a CW direction as far as it could go. Other limit switches would be mounted on the shoulder and elbow. They would close when the respective joint was fully extended. Another switch mounted to the wrist would close when the wrist was rotated to the furthest CW position. The last switch mounted to the gripper would close when the gripper is fully opened. To reset the arm to its home position, each function is activated in the direction of travel needed to close a limit switch until that limit switch actually closed. After all the functions are in the home position, the computer would then know the absolute position of the arm.

Once home, we can start and replay a script, knowing that the drift occurring during one run of the script is probably so minimal that the robotic arm will hit all its marks. Once the script is finished, the arm is reset to its home position and the script file is replayed.

Sometimes a home position does not give enough feedback to perform certain operations, for instance, picking up an egg without crushing the shell. For these applications, more sophisticated methods of feedback need to be employed. Signals from transducers are processed using analog-to-digital (A/D) converters. The processed signals are used to determine factors such as position, pressure, speed, and torque. A simple example will illustrate. Imagine mounting a small linear potentiometer on the gripper section. The potentiometer is mounted so that when the gripper opens or closes, the slider on the potentiometer slides back and forth. So as the gripper opens and closes, the resistance of the potentiometer varies. Once calibrated, the resistance could accurately tell how far the gripper closed (or opened).

Feedback systems add another layer of complexity and cost to the system. One can always use the manual control system to override and reposition the robotic arm as a script is running.

## Connecting manual control to interface

After the interface is running properly, connect manual control to the interface using the eight-pin header. Orient the manual control's Molex connector to the eight-pin interface header as shown in Fig. 15.10. Press the connector firmly onto the header to seat. The

■ **15.10** *Connecting the manual control*

robotic arm can be controlled manually at any time. It doesn't make a difference if the interface is connected to the computer or not.

## DOS-level keyboard program

The keyboard program is a DOS-level program that allows one to control the robotic arm in real time (interactively) using the keyboard. The following keys perform the following functions:

| Key | Function |
| --- | --- |
| U | Up |
| L | Left |
| G | Grip |
| S | Stop |
| D | Down |
| R | Right |
| H | Release |
| Q | Quit |

## Speech control for robotic arm

The speech control for the robotic arm uses the speech-recognition kit from Chap. 7. In this section we will build an interface from the speech-recognition kit to the robotic arm. This interface is also offered in a kit form from Images SI, Inc.

The schematic for the speech-recognition interface is shown in Fig. 15.11. The interface uses a 16F84 microcontroller. The program for the microcontroller is as follows:

```
' Speech Recognition Interface program
Symbol PortA = 5
Symbol TRISA = 133
Symbol PortB = 6
Symbol TRISB = 134
Poke TRISA, 255
Poke TRISB, 240
Start:
Peek PortB, B0
If bit4 = 0 then trigger     'Trigger enabled, read speech-
                             recognition circuit
goto start    'Repeat
trigger:
Pause 500    'Wait 0.5 seconds
Peek PortB, B0    'Read BCD number
```

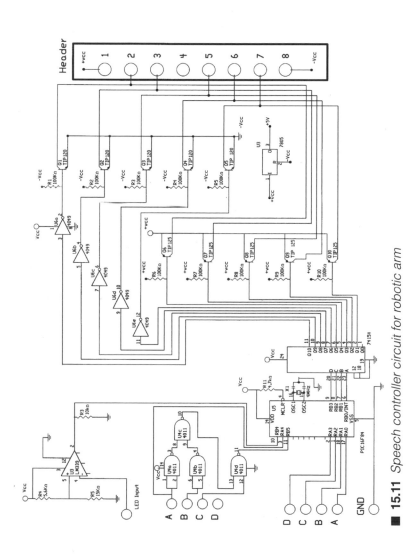

**15.11** Speech controller circuit for robotic arm

```
if bit5 = 1 then send     'Output number
goto start    'Repeat
send:
peek PortA,b0    'Read port A
if bit4 = 1 then eleven     'Is the number 11?
poke PortB, b0    'Output number
goto start    'Repeat
eleven:
if bit0 = 0 then ten
poke portb,11
goto start    'Repeat
ten:
poke partb,10
goto start    'Repeat
end
```

Updates to the 16F84 program may be downloaded for free at
http://www.imagesco.com.

## Programming the speech-recognition interface

Programming the speech-recognition interface is the same procedure used to program the speech-recognition kit in Chap. 7. To
operate the robotic arm properly, you must program certain word
numbers to specific robotic arm functions. You may use whatever
command word you wish for any particular command. I am providing a command word in Table 15.1 for illustration; you may
change any word you wish.

■ Table 15.1

| Word Number | Typical Command Word | Robotic Arm Function |
|:---:|:---:|:---|
| 1 | GRIP | Close gripper |
| 2 | E-Down | Elbow down |
| 3 | R-Base | Rotate base CCW |
| 4 | S-Up | Shoulder up |
| 5 | L-Wrist | Rotate wrist CW |
| 6 | Release | Open gripper |
| 7 | E-Up | Elbow up |
| 8 | L-Base | Rotate base CW |
| 9 | S-Down | Shoulder down |
| 10 | R-Wrist | Rotate wrist CCW |
| 11 | Stop | Stop |

## Parts list for the PC interface

- ☐ (5) Tip 120 NPN transistors
- ☐ (5) TIP 125 PNP transistors
- ☐ (1) 74164 serial to parallel IC
- ☐ (1) 74LS373 octal latch
- ☐ (1) Red LED
- ☐ (5) 1N914 diodes
- ☐ (1) Eight-position Molex header
- ☐ (1) Eight-position 3"-long Molex cable
- ☐ (1) DPDT PC-mounted switch
- ☐ (1) DB25 RT angled PC-mounted connector
- ☐ (1) DB25 M-M 6-ft cable
- ☐ (1) PCB
- ☐ (10) 100K-ohm, $\frac{1}{4}$-W resistors
- ☐ (3) 15K-ohm, $\frac{1}{4}$-W resistors
- ☐ (1) 7805 voltage regulator

The robotic arm interface kit contains all the above parts.

## Parts list for the speech-recognition interface

- ☐ (5) TIP 120 NPN transistors
- ☐ (5) TIP 125 PNP transistors
- ☐ (1) 74154 4/16 decoder IC
- ☐ (1) 4011 NAND gate
- ☐ (1) 4049 hex buffer
- ☐ (1) 741 op-amp
- ☐ (1) 5.6K-ohm, $\frac{1}{4}$-W resistor
- ☐ (1) 15K-ohm, $\frac{1}{4}$-W resistor
- ☐ (1) Eight-position Molex header
- ☐ (1) Eight-position 3"-long Molex cable
- ☐ (1) PCB
- ☐ (10) 100K-ohm, $\frac{1}{4}$-W resistors
- ☐ (1) 4.7K-ohm, $\frac{1}{4}$-W resistor
- ☐ (1) 7805 voltage regulator

- ☐ (1) PIC16F84 microcontroller
- ☐ (1) 4.0-MHz crystal
- ☐ Robotic arm interface kit—$44.95
- ☐ OWI robotic arm trainer—$84.95
- ☐ Speech-recognition interface to robotic arm—$39.95
- ☐ Speech-recognition kit—$100.00

Parts are available from:

Images SI, Inc.
39 Seneca Loop
Staten Island, NY 10314
(718) 698-8305

# 16

# Android hand

IN THIS CHAPTER WE WILL CONSTRUCT A HUMANLIKE OR android hand. The actuator we will use to move the fingers in the android hand is the air muscle introduced in Chap. 3.

The air muscle is a pneumatic device that produces linear motion with the application of pressurized air. Much like a human muscle, it contracts when activated. You may think, well, pneumatic cylinders have been around quite a while and do the same thing. True, but the air muscle represents a boon to hobbyists and robotists because it is lower in cost, extremely lightweight, flexible, and easier to use.

The air muscle has a power-to-weight ratio of about 400:1. Since most of its components are plastic and rubber, the air muscle can work while wet or underwater. The flexible nature of the air muscle allows it to be connected to and contract on/off axis pulleys and levers. The air muscle can contract even when bent around curved surfaces. These easy-to-use features make the air muscle the experimenter's choice over standard pneumatic cylinders.

Of course, being a pneumatic device it needs a supply of compressed air. Compressed air is not as readily available as electric current. When I first learned about the air muscle, I thought that a small air system would be too much of a hassle to build. I was wrong. A simple air system can be put together for about $25.00, and a small electric air system for about $50.00.

Overall efficiency is lost when using electric power to compress air. However, the air muscle consumes little air volume per activation and the compressed air can be stored. The air muscle's response and cycle times are fast. A small 6" 10-gram (g) air muscle can lift 6.5 pounds (lb).

Before we build the android hand, we will first build a few manually operated air muscle demonstration devices. The demonstration devices allow you to become familiar with the operation and function of the air muscle, before attempting a more complex project.

Manual control of an air muscle is fine for projects needing one or two air muscles. However, when five or six air muscles need to be operated in sequence or unison, manual control is not practical. Instead, we employ computer control. One may use an IBM PC or compatible PIC microcontroller. The interface to either computer is the same. In this chapter we will use the IBM PC. To control the air muscle via a computer (IBM or compatible printer port) through the PC's parallel port adds approximately $25.00 per air muscle to the cost.

## Advantages of the air muscle

☐ *Light weight.* Six-inch air muscle with 18" of $5/_{32}$"-diameter air tubing weighs approximately 10 g.

☐ *Contraction.* Six-inch air muscle contracts approximately 1" (about 25 percent of its length, ends not included).

☐ *Powerful.* Generates a force of approximately 6.5 lb at 42 pounds per square inch (psi). The power-to-weight ratio can reach 400:1.

☐ *Pliable.* Soft, pliable construction and can be bent around curved surfaces and still function properly.

### Uses

The air muscle lends itself to robotics and automation. In some applications it can replace servo motors and direct current (DC) motors. Its unique properties—lightweight, strong, and pliable—can be capitalized on in many applications and used to improve existing pneumatic designs. In a nutshell, the air muscle may be used in many applications that require linear or contractive motion. In many cases pneumatic cylinders can be replaced.

### How the air muscle works

The air muscle has a long tube constructed out of black plastic mesh. Inside of it is a soft rubber tube. Metal clips are fastened on each end. The plastic mesh is formed into loops on each end, tucked into and secured by the metal clips. The loops are used for fastening the air muscle to devices.

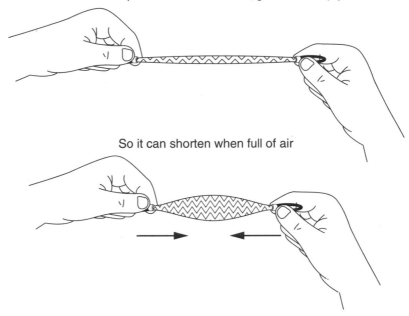

You need to pull the muscle out straight when empty

So it can shorten when full of air

■ **16.1** *Drawing of air muscle being stretched*

When air pressure is applied, the muscle contracts by the following mechanism. When the air muscle is pressurized, the soft inner tube expands. The inner tube pushes against the black plastic mesh and causes it to expand also. When the plastic mesh expands, it shortens in length in proportion to the expansion of its diameter. This causes the air muscle to contract. However, it is essential that the air muscle be in a stretched or elongated position when it's deactivated or in a resting state in order for it to operate properly. If not, there will not be any movement or contraction when it is activated (see Fig. 16.1).

## Components of the air muscle system

Figure 16.2 illustrates the components needed to use the air muscle. Item 1 is the air muscle itself (of course). Item 2 is a three-way air valve. The three-way air valve allows one to manually operate the air muscle (see Fig. 16.3).

Item 3 is a bottle top adapter, with a pressure release valve (set around 60 psi). The bottle top adapter allows one to use a standard plastic PET soda bottle for air storage. The pressure release valve automatically releases excess air when preset pressure is exceeded.

1) One standard 6" air muscle

2) Three-way valve for controlling the air flow

3) Bottle cap adapter for attaching an air storage

4) PET soda drink bottle to be used as the air storage

5) Foot pump adapter for connecting a foot pump to the $5/32$" air line

6) Foot pump

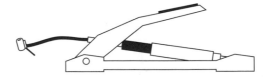

7) Nylon cable ties, to attach the air muscle to your device

■ **16.2** *Items needed to experiment with the air muscle*

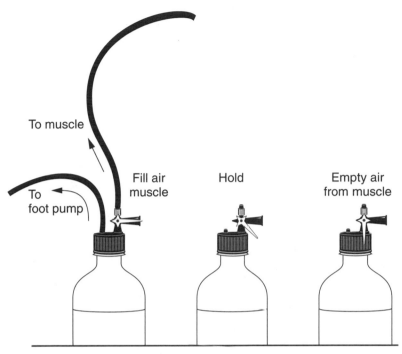

To muscle

To foot pump

Fill air muscle

Hold

Empty air from muscle

■ **16.3** *Three-way air valve for air muscle operation*

Item 4, the PET soda bottle, is used for air storage. A plastic soda bottle can hold 50 psi easily. I have static-tested plastic PET bottles to 100 psi. *Caution: Never use any type of glass bottle for air storage.* A slight fracture in a glass bottle or dropping it accidentally may cause the bottle to explode, sending tiny glass fragments all over. Plastic PET bottles elongate when overpressurized.

Item 5 is the foot-pump adapter and item 6 is the foot pump. A simple foot pump with an air pressure gauge can charge air storage up to 100 psi. Because of the low volume of the PET bottles, air storage is brought to 50 psi with three or four strokes of the air pump. The air muscle uses so little air that a small PET bottle holds enough air for four to five complete cycles of the air muscle. Item 7 is nylon cable ties, used to quickly connect the air muscle to a mechanical device.

Figure 16.4 illustrates a general overview of how the parts are put together. In some cases you may want to epoxy glue some components together to prevent them from popping off. For instance, if you will just be using the three-way air valve on one bottle adapter for air muscle experiments, you may want to glue the three-way valve to the adapter permanently.

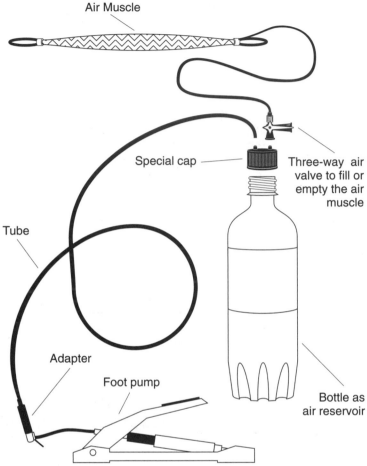

This is how they all fit together...

Air Muscle

Special cap

Three-way air valve to fill or empty the air muscle

Tube

Adapter

Foot pump

Bottle as air reservoir

■ **16.4** *General overview of how parts fit together*

### Attaching the air muscle to mechanical devices

The air muscle is made of a soft inner tube encased in a strong plastic mesh. The assembly is held together by metal clips on each end. The plastic mesh is looped at each end making a hole. The plastic mesh hole is very strong mechanically and can be used to attach the air muscle to any device. Figure 16.5 shows a machine screw inserted through the mesh hole.

### Using the air pump adapter

When you receive your foot pump, it will have a standard air nozzle as shown in Fig. 16.6. We need to replace the standard nozzle with the air pump adapter. Lift the locking lever as shown in Fig. 16.7.

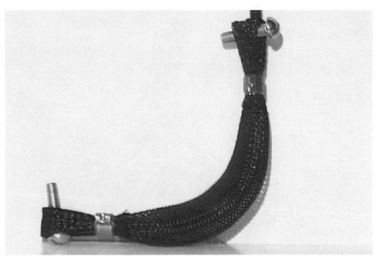

■ **16.5** *Screw going through one end loop of air muscle*

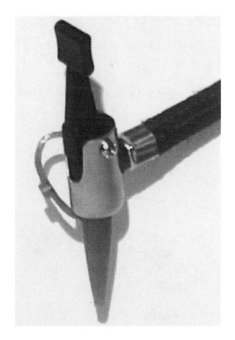

■ **16.6** *Foot-pump nozzle*

Remove the standard nozzle (see Fig. 16.8), and insert the air pump adapter (see Fig. 16.9). Close the locking lever by pressing it back down.

## Have a Coke or Pepsi

You need to acquire a plastic PET bottle. The easiest way to do so is to buy a soda. Make sure the soda bottle is plastic. Don't purchase a

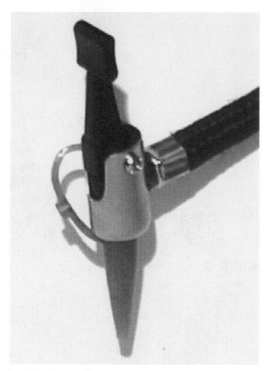

■ **16.7** *Lift locking lever (foot-pump nozzle)*

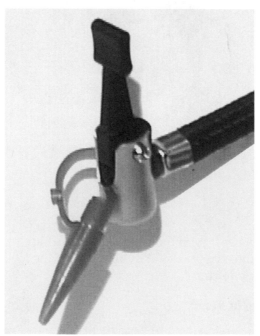

■ **16.8** *Remove standard nozzle adapter*

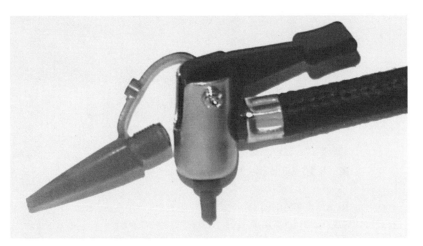

■ **16.9** *Insert air foot-pump adapter*

soda bottle larger than 1 liter (L). A half-liter bottle is ideal. I've tried the bottle cap adapter on all sizes of PET soda bottles up to 2 L, and it fits all of them.

Empty the PET bottle of soda and clean it out with tap water. The bottle should be completely dry before using. It's interesting to note that if a full bottle of soda is dropped, the resulting pressure caused by the release of the carbonated soda greatly exceeds the 50-psi limit we impose on the bottle. Soda companies have designed PET bottles to withstand a rapid increase in bottle pressure, which would come from dropping the bottle. This is something I never realized before working with the air muscle, and thought I would pass it on. Remember, no glass bottles should be used in the air muscle pneumatic system.

## Building the first demo device

The first mechanical device we will build is a simple one that can be used to measure the contraction of the air muscle (see Fig. 16.10). The base is 1" × 2" lumber and is 11" long. I used this material simply because I had it lying around. You can just as easily use metal or plastic. At each end I drilled a hole to accept a $1^{3}/_{4}$"-long 8-32 machine screw. The machine screws are inserted and held in place using two 8-32 nuts, one nut on each side of the wood. The head of the screw and shaft protrude about $^{3}/_{4}$" above the wood.

The top screw is threaded through the top opening of the air muscle, before inserting the screw into the wood. A rubber band is looped through the bottom opening of the air muscle and then

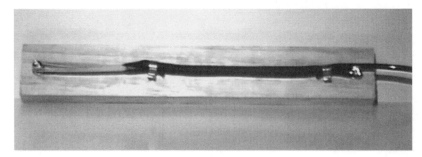

■ **16.10** *First mechanical device*

looped around the bottom screw. The rubber band stretches the air muscle when it is in its relaxed state.

Connect the balance of components as shown in Fig. 16.4. At times I've had difficulty connecting the $^5/_{32}$"-diameter tubing to some of the components. Here are a few tips. First, if the tube refuses to go onto an adapter, place the tubing under running hot water from the faucet. This softens the plastic, making it easier to fit onto the components. Another trick is to use some clear plastic tubing. The plastic tubing is snug enough to fit onto the adapter nozzles properly (see Fig. 16.11). In addition it is pliable enough to fit the $^5/_{32}$"-diameter tubing inside the tubing itself (see Fig. 16.12). The soft tubing acts like an adapter and quick release for changing air muscle devices.

To operate the device, first pressurize the system using the foot pump. It only takes about four strokes to reach 50 psi. Your mileage may vary depending upon the size of the PET bottle you are using.

Open the three-way air valve to charge the air muscle. The muscle should immediately contract. You can measure the distance it moves in proportion to the psi gauge on the pump. You should be able to operate the muscle through four or five contractions and expansions before you need to refill the PET bottle. The air muscle doesn't use much air.

Notice that the air muscle stays in the contracted position until the three-way valve is turned to release the air from the muscle. It doesn't cost any energy to keep the air muscle contracted. This is in contrast to servo motors and solenoids that must be supplied electric energy continuously to maintain their push or pull.

If the muscle doesn't appear to contract, then it probably wasn't stretched far enough. Remember the muscle must be stretched in order for it to contract (operate).

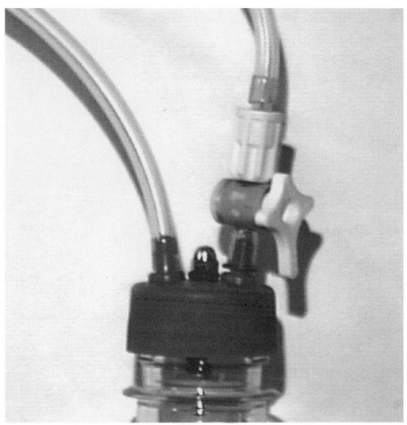

■ **16.11** *Using clear plastic tubing on standard adapters*

■ **16.12** *Using clear plastic tubing and $5/32$" tubing*

## Building the second mechanical device

The second device is a lever (see Figs. 16.13 and 16.14). The lever I made is constructed out of wood and plastic. Machine screws secure the air muscle and rubber bands to the lever arm. A wood screw through the plastic arm is the pivot. A second wood screw holds both the air muscle and rubber band. Operate this device using the three-way air valve as before. When activated, the lever moves up.

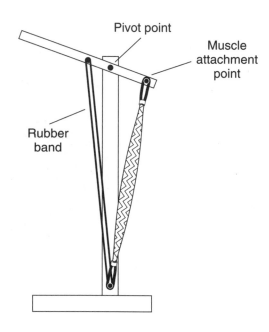

When the air muscle fills, it shortens, pulling the lever up.

When the air is let out, the muscle lengthens and the elastic bands pull the lever down.

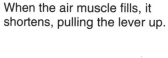

■ **16.13** *Second mechanical device "lever"*

■ **16.14** *Second mechanical device "lever"*

## IBM interface

Computer control is easy. The computer operates an electric three-way air valve. An inexpensive three-way electrically operated solenoid air valve is available (see Fig. 16.15). This air valve operates at 5 volts DC (VDC) and is rated at 90 psi. This air valve has quick connect and disconnect air ports. The $^{5}/_{32}$"-diameter stiff tubing is simply inserted into the port hole, and it locks in. To disconnect, hold and secure the port ring with your fingers into the air valve and tug the $^{5}/_{32}$"-diameter air tubing out.

To operate a single air valve, we only need one pin off the parallel (printer) port, along with a ground (see Fig. 16.16). The output pin is buffered with a gate off of a 4050HCT noninverting hex buffer. The output of the hex buffer turns a TIP 120 NPN Darlington transistor on or off. The transistor controls the current going to the air valve.

■ **16.15** *Electrically operated three-way air valve*

## BASIC program

The BASIC program is short and simple. After finding the printer port address, the following lines control the valve of pin 2:

```
5 REM Solenoid Air Valve Controller
10 REM John Iovine
15 REM Find Printer Port Address
20 DEF SEG = 0
25 a = (PEEK(1032) + 256 * PEEK(1033))
30 REM Next line activates the air muscle
35 OUT a,1
40 REM Next line deactivates the air muscle
45 OUT a,0
```

By bringing the DB25 pin 2 high, the air valve is opened allowing air pressure to the air muscle. Bringing pin 2 low, closes the air valve to the muscle and vents the air from the air muscle.

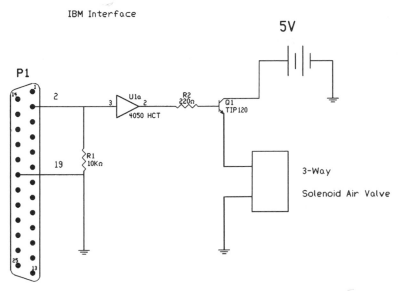

IBM Interface

P1

2

3   U1a   2
        4050 HCT

R2
220Ω

Q1
TIP 120

5V

R1
10KΩ

19

3-Way

Solenoid Air Valve

DB25 Connector
To Printer Port

■ **16.16** *Schematic of air valve controller*

## More air

The air muscle, as previously discussed, uses compressed air from a PET plastic air storage bottle and foot pump. One can use compressed air from just about any source that's available. For instance, you can purchase small compressed air canisters used for air brushing. In fact, airbrushing supplies may provide you with a list of suitable tubing and fittings to experiment with.

There are a few small electric air compressors available on the market. The more expensive ones, made for airbrush painting, have metal storage containers and air pressure regulators. At the other end of the market are the inexpensive 12-VDC portable air compressors used for tire inflation. These compressors typically do not have an air pressure regulator or air storage. These items may be purchased to make an inexpensive pneumatic system.

Never use plastic PET bottles for air storage with any kind of automatic air compressor. The plastic PET bottles are only suitable for air storage with hand (or foot) operated air pumps. Always use an air storage tank with any automatic air compressor. Small air storage tanks are not expensive.

### Safety first

Since this is a new product, not too many people may be familiar with working with pneumatic systems. Therefore, a few safety guidelines should be followed.

1. Always wear goggles when prototyping a new design.
2. Never connect a plastic PET soda bottle to an air compressor.
3. Never use a glass bottle for air storage.
4. Limit PET bottle size to 1 L (or quart) or less.
5. Do not unscrew the bottle top or pull off an air fitting or valve when the system is still pressurized. Bleed the system of air first.

## Android hand

The construction of a human-type hand, grasping mechanism begins with a trip to a toy store. The toy needed is called an Awesome Arm, made by Zima Company in China (see Fig. 16.17). You will need to purchase two Awesome Arms to get enough fingers. The thumb on the toy is fixed and cannot be used.

The toy works by allowing the users to use their own fingers to actuate fingers on a robotic hand, something like a hand or arm extension. We will scavenge the main component out of this toy to make an inexpensive android hand.

When you turn the arm over, there are five small screws that hold the hand together. Remove the screws and the arm section will come apart (see Fig. 16.18). Remove the finger section of the toy (see Fig. 16.19). Discard the rest of the components. The rectangular boxes on the finger pulls are where people place their fingers to use the extension hand. We won't be needing them, so cut them off using wire cutters, leaving a long plastic stem.

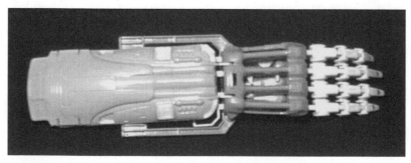

■ **16.17** *Awesome arm manufactured by Zima Company*

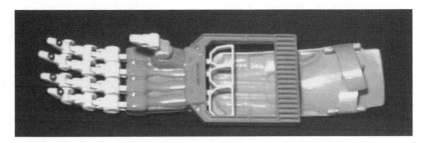

■ **16.18** *Opposite side of arm, where screws are removed*

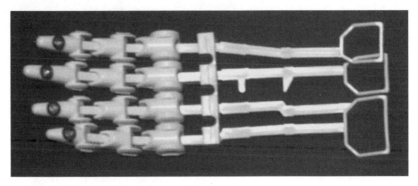

■ **16.19** *Finger pulls salvaged from arm*

**315**

We need to build a substructure to support all the components. I began by tracing the outline of my right hand on paper. Then I shaded in a form that would become the support structure (see Fig. 16.20). The shaded area was cut from $1/8$"-thick aluminum plate.

The fingers must be secured to the end of the plate. To do so, first mark the position on the support aluminum. Next, place a small aluminum plate $1/2$" wide $\times$ $1/8$" thick right behind the plastic back of the fingers (see Fig. 16.21). This forms a back plate for the finger base to rest against. Drill three holes through the two aluminum plates, and fasten the small plate in position using a few machine screws and nuts.

Secure a top aluminum plate $1/8$" $\times$ $1/2$" to the top of the finger base. Drill four holes through the top plate and support plate as shown by the screw positions in Fig. 16.22. Four 1"-long machine screws and nuts secure the top plate in position. These machine screws serve a dual purpose. First, they secure the top plate that locks the fingers onto the support plate. Second, they will hold a rubber band to provide tension for the air muscle.

With the fingers secured to the support plate, we need to attach an air muscle to each finger. In order for the air muscle to provide a

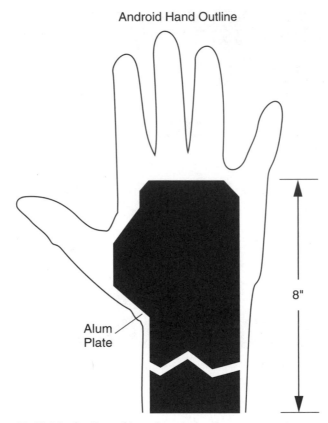

Android Hand Outline

Alum Plate

8"

■ **16.20** *Outline of hand and aluminum support*

useful contraction, it must be stretched. Loop a rubber band through the air muscle. Then remove the first of the four 1"-long screws securing the top plate. Place the looped ends of the rubber band where the screw passes through the top plate. Replace the screw, threading it through the looped ends and through the top plate holes, and then secure it with a nut (see Figs. 16.23 and 16.24).

Pull back the air muscle until it's fully extended. Mark the end of the air muscle. Drill a hole on the mark, and place the machine screw and nut there. Place the end loop of the air muscle over the machine screw to hold the air muscle in an extended position (see Fig. 16.25).

Now drill a small hole in the plastic section of the finger pull. The small hole should line up with the front loop of the air muscle. The hole must just be large enough to allow a double strand of magnetic wire to pass through it. You can substitute bare 22-gauge solid copper wire for magnetic wire.

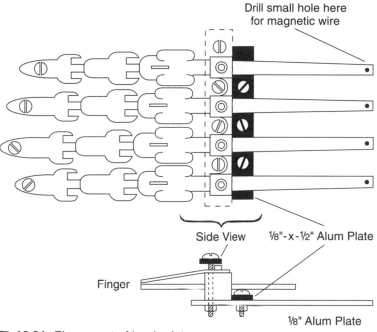

Drill small hole here
for magnetic wire

Side View    ⅛"-x-½" Alum Plate

Finger

⅛" Alum Plate

■ **16.21** *Placement of back plate*

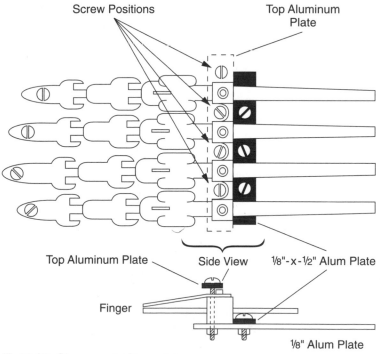

Screw Positions    Top Aluminum
Plate

Top Aluminum Plate    Side View    ⅛"-x-½" Alum Plate

Finger

⅛" Alum Plate

■ **16.22** *Placement of top plate*

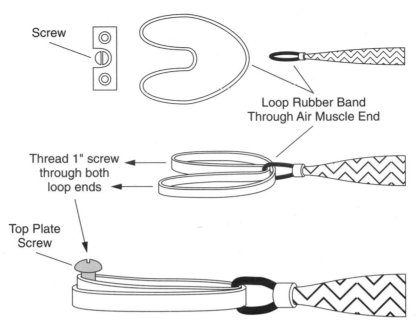

Screw

Loop Rubber Band
Through Air Muscle End

Thread 1" screw
through both
loop ends

Top Plate
Screw

■ **16.23** *Threading rubber band through one end of air muscle and attaching opposite end to top plate screw*

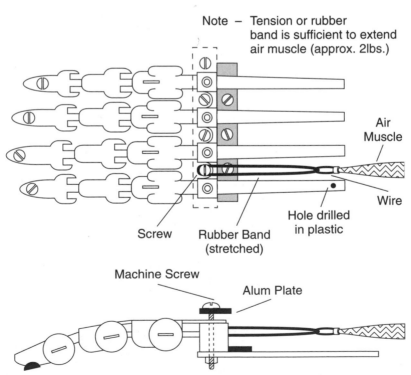

Note – Tension or rubber
band is sufficient to extend
air muscle (approx. 2lbs.)

Air
Muscle

Wire

Hole drilled
in plastic

Screw

Rubber Band
(stretched)

Machine Screw

Alum Plate

■ **16.24** *Overview of attaching stretched air muscles to finger pull*

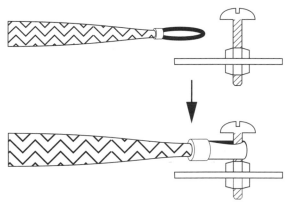

■ **16.25** *Attaching opposite end of air muscle to machine screw to extend air muscle*

Pass a double strand of wire through the plastic hole and front loop of the air muscle. Twist the ends of the wire together securing the components together. If there is excessive wire left from twisting, clip it off using wire cutters.

The top view should look something like Fig. 16.24. We can now see how the finger will contract. As the air muscle is pressurized, it contracts. The contraction pulls the plastic stem of the finger pull, which in turn contracts the finger. When pressure from the air muscle is released, the rubber band extends the air muscle back into its original extended position.

At this point it's a good idea to static test the finger. Connect the air supply to the muscle to ensure it operates in the manner just described. The prototype required a pressure of 42 psi to fully contract the index finger.

When the finger operates properly, connect the air muscles to the remaining fingers in the same manner described. Figure 16.26 is a close-up of the air muscles connected to all the finger pulls.

## The thumb

The thumb is the most important finger on the hand. It makes grasping, holding, and using tools much easier. Don't think so? Try picking up a coin off a table or floor without using your thumb. Now try using a few tools, like pliers, wire cutter, hammer, or drill.

To make the thumb, cut off the small finger assembly from the second hand unit purchased. Assemble this finger section lower and at a 45 degree angle to the other fingers (see Fig. 16.27).

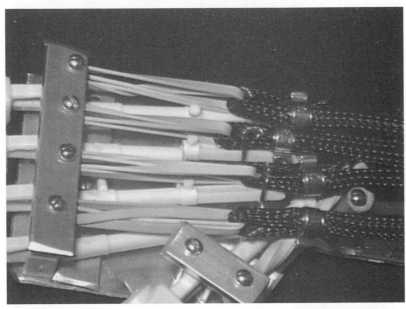

■ **16.26** *Close-up of air muscle, rubber band, and finger pull tied together in finished hand*

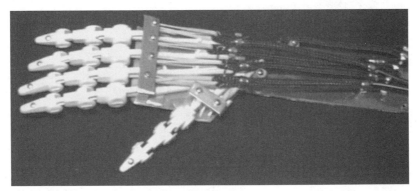

■ **16.27** *Finished robotic hand*

The thumb in this prototype is articulated (moves) but is not opposable. To improve this design, make the thumb opposable. This will increase the effectiveness of the hand. To make the thumb opposable, the thumb-containing portion of the hand must be cut off and reattached using a spring-loaded hinge (see Fig. 16.28). The spring-loaded hinge would be located on the rectangular box shown in Fig. 16.28. An air muscle connects to this section; when activated, it pulls the thumb into the palm section of the hand. This makes the thumb opposable as well as articulated.

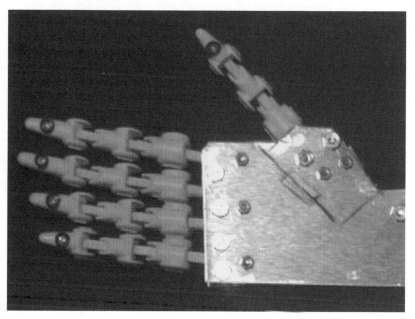

■ **16.28** *Plans to make thumb opposable as well as articulated*

## Going further

The robotic hand can be interfaced to an IBM-compatible computer using five electric solenoid valves, similar to the single-valve design shown earlier. An outer covering like a rubber hand can be fitted over the robotic hand to make an android hand (see Fig. 16.29).

Some other applications for the air muscle are interesting. Here are a few:

☐ Six-legged robotic walker

☐ Easy-open jar clamp (for people with arthritis)

☐ Robotic hands

☐ Robotic arms

## Parts list for the air muscle

☐ (1) Air muscle, 6" long with $5/32$"-diameter tubing—$15.95

☐ (1) PET bottle top adapter with pressure release valve—$4.00

☐ (1) Three-way air valve—$4.00

☐ (1) Air pump adapter—$2.00

☐ (1) Foot air pump with 100-psi air pressure gauge—$12.95

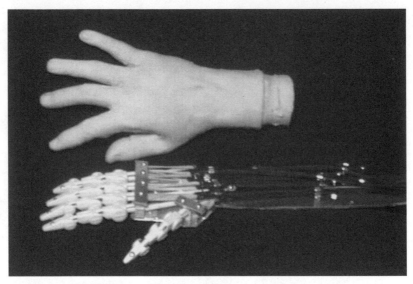

■ **16.29** *Fitting lifelike rubber hand over robotic hand to create an android hand*

- ☐ (1) $5/32$"-diameter air tubing—$0.25 per ft
- ☐ (1) $7/32$"-diameter clear air tubing (for making quick releases)—$0.25 per ft

## Parts list for the IBM interface

- ☐ (1) 5-VDC three-way solenoid air valve, 90 psi maximum—$30.00
- ☐ (1) DB25 pin connector—$3.50
- ☐ (1) 4050HCT noninverting hex buffer—$4.00
- ☐ (1) TIP 120 NPN Darlington transistor—$1.25

Parts are available from:

Images Company
39 Seneca Loop
Staten Island, NY 10314
(718) 698-8305

`http://www.imagesco.com`

# Suppliers

Jameco Electronics
1355 Shoreway Road
Belmont, CA 94002
(650) 592-8097

JDR Electronics
1850 South 10 Street
San Jose, CA 95112
(800) 538-5000

Images SI, Inc.
39 Seneca Loop
Staten Island, NY 10314
(718) 698-8305

Radio Shack
Check local telephone directory for store nearest you

# Index

*327*

# About the Author

John Iovine is the author of several popular McGraw-Hill titles that explore the frontiers of scientific research. He has written *Homemade Holograms: The Complete Guide to Inexpensive, Do-It-Yourself Holography*; *Kirlian Photography: A Hands-On Guide*; *Fantastic Electronics: Build Your Own Negative-Ion Generator and Other Projects*; and *A Step into Virtual Reality*. He is also the "Amazing Science" columnist for *Poptronics* magazine.